CREATIVE CONSTRUCTION OF MATHEMATICS AND SCIENCE CONCEPTS IN EARLY CHILDHOOD

by Nancy L. Gallenstein

Association for Childhood Education International
17904 Georgia Ave., Ste. 215, Olney, MD 20832
800-423-3563 • www.acei.org

Anne W. Bauer, ACEI Editor
Bruce Herzig, ACEI Editor

Copyright © 2003, Association for Childhood Education International
17904 Georgia Ave., Ste. 215, Olney, MD 20832

Library of Congress Cataloging-in-Publication Data
Gallenstein, Nancy L.
    Creative construction of mathematics and science concepts in early childhood / Nancy L. Gallenstein.
       p. cm.
Includes bibliographical references.
   ISBN 0-87173-159-2 (pbk.)
  1. Mathematics—Study and teaching (Early childhood) 2. Science—Study and teaching (Early childhood) 3. Critical thinking in children. I. Title.
   QA135.6 .G35 2003
   372.7—dc21

                                                             2002011581

Table of Contents

5   **Introduction**

9   **Chapter One**
Children's Acquisition of Mathematics and Science Knowledge

31   **Chapter Two**
Making Learning Meaningful Through the Learning Cycle Lesson Format

51   **Chapter Three**
Concept Attainment Methods for Early Childhood Learners

81   **Chapter Four**
Concept Mapping for Early Childhood Learners

93   **Chapter Five**
Tying It Together

Introduction

Children at the early childhood level (ages 3-8) are active participants as they learn fundamental concepts and process skills. Children develop their understanding of concepts as they grow, and each child's rate of development differs. Many concepts that children acquire as they develop are basic to both mathematics and science, such as one-to-one correspondence, number, patterning, sets, comparison, classification, communication, counting, measurement, and graphing. Furthermore, many math and science concepts are interrelated, as fundamental math concepts are necessary when solving problems in science. As children confront new situations through exploration, observing their environment through play (with blocks, water, sand, etc.), they apply these process skills and basic math and science concepts as they collect and organize information. These basic concepts acquired at the early childhood level are applied later when exploring more abstract concepts in mathematics and science.

Concepts are acquired when they are understood. Children are most likely to reach that understanding through meaningful learning experiences that emphasize active, hands-on conceptual learning, rather than through presentation of isolated facts and procedures—as is often the case in mathematics and science curricula (Charlesworth & Lind, 2003). Children should be presented with relevant opportunities to think critically, problem solve, and make decisions as they interact with a variety of age-appropriate materials, and with their peers and informed adults. This text presents some effective models for promoting such vital life skills.

Effective teaching models that emphasize the importance of critical thinking in mathematics and science are often integrated into intermediate, middle school, and high school curricula, but are less often used in early childhood classrooms. By utilizing Jean Piaget's cognitive learning theory (1954), Jerome Bruner's discovery learning theory (1967), and Lev Vygotsky's social interaction theory (1987), however, specific teaching models that are effective with more mature learners can be adapted to promote critical thinking for children ages 3-8. Children at all learning levels should have opportunities to develop and expand their critical thinking skills while creatively constructing mathematics and science concepts.

This book provides preservice and inservice teachers with an explanation of teaching models that promote critical thinking, problem solving, decision making, and cooperative learning through the creative construction of mathematics and science concepts. The author also presents ways that early childhood education mathematics and science methods instructors can teach and promote these creative models on the university/college level. Although the models presented in this book may not be new to many educators, the suggested adaptations for early childhood may be. As teachers experiment with these models, they will discover additional, novel ways to incorporate them into their early childhood curricula.

Personal Perspective on the Value of Critical Thinking

Promoting students' critical thinking as they construct their math and science knowledge is important for many reasons. One reason is that the majority of early childhood and elementary teachers have spent most of their lives in educational settings, and so they have limited work experiences outside the field of education. The majority of the students they teach, however, *will* seek and acquire jobs outside of educational settings.

From the corporate perspective, the most valuable employees are those who possess both social and intellectual competencies. Employees are expected to work well with others, think critically and creatively, solve problems, and make effective decisions. Whether businesses stay afloat or sink is directly attributable to their employees' skills in these areas.

Teachers must not only provide their students with opportunities to engage in mathematics and science activities that promote critical thinking, but also value their role in this process. Children deserve such opportunities, and the future depends on educators' ability to prepare children well in this area.

Overview of Critical Thinking Models and Connection to Constructivism

Various teaching/learning models exist that promote children's construction of mathematics and science knowledge through critical thinking. For example, the Learning Cycle model, which was fueled by Piaget's constructivist theory and credited to Robert Karplus (1967) and his team of educators in the 1960s, is used in the Science Curriculum Improvement Study (SCIS). This teaching model has been exceptionally effective in terms of students' retention of science concepts (Martin, Sexton, & Gerlovich, 2001). While participating in this model, children progress through five phases. During the initial phases, children's prior knowledge is tapped. Children then build on their prior knowledge by discovering and reinventing knowledge through various explorations. Eventually, children have opportunities to experience how real life, including careers, is related to their discovered/reinvented concepts. Educators can, in addition to teaching science concepts, promote this model when teaching mathematics concepts to early childhood-age learners.

Another critical thinking teaching model that is often used with older children and can be adapted for young children by incorporating concrete experiences is Jerome Bruner's (1967) approach to attaining concepts through critical thinking and discovery. This inductive method provides children with opportunities to draw on their prior knowledge while comparing and contrasting (classification is emphasized in both early childhood math and science) information provided by the facilitator of each concept attainment learning experience (Joyce, Weil, & Calhoun, 2000).

Joseph Novak's (1984) Concept Mapping technique is another effective model; it is based on the theory of David Ausubel (1963), who also emphasized the importance of tying prior knowledge to new concepts, as well as the value of organizing knowledge in a graphic format. Through the concrete/visual graphic representations promoted in this model, children's misconceptions about mathematics and science often can be clarified and addressed. Concept Mapping also serves as an excellent pre- and post-assessment tool for teachers when evaluating children's conceptual understandings.

The above-mentioned models challenge children to build on their prior knowledge by extending their thinking and creating links to newer learning in both mathematics and science, while taking ownership of newly acquired concepts.

Overview of Book

The goal of this text is to present educators with effective teaching/learning models that will challenge early childhood-age children to creatively construct mathematics and science concepts.

Chapter One contains a review of information about the developmental levels of children in relation to mathematics and science learning, as well as an examination of how children acquire mathematics and science knowledge. Appropriate math and science concepts as well as science process skills for early childhood-age children are also discussed, with an emphasis on the value of critical thinking as children construct knowledge. Then, readers will learn about specific techniques for integrating mathematics and science concepts during early childhood instruction.

Chapter Two includes an adapted Learning Cycle Lesson format (5E - Engagement, Exploration, Explanation, Expansion, and Evaluation) for children, and provides a historical review of the original model. Thorough examples of mathematics and science Learning Cycle lessons include suggestions for implementation. The numerous benefits of the Learning Cycle lesson format are outlined.

Chapter Three includes a discussion of Jerome Bruner's emphasis on the importance of children attaining concepts through inductive reasoning. The examples provided demonstrate early childhood mathematics and science concept attainment activities for whole-group instruction. In addition, various adaptations of the presented concept attainment model include examples for implementation. These variations can be used for whole-class instruction or they can be placed in learning centers or on interactive bulletin boards. Early childhood educators can create their own concept attainment activities by following the original model or the suggested adaptations. The chapter concludes by detailing the benefits of incorporating concept attainment activities into the early childhood mathematics and science curriculum.

Chapter Four introduces the Concept Mapping technique, provides a discussion of the format for implementation with early childhood-age children, and explains the technique's benefits. Examples of concept mapping for both early childhood mathematics and science concepts include suggestions for implementation.

Chapter Five summarizes the critical thinking models for early childhood-age children that are included in this book. An emphasis is placed on the overall benefits of the presented teaching/learning models, and on the need to incorporate critical thinking into the early childhood mathematics and science curriculum.

References

Ausubel, D. P. (1963). *The psychology of meaningful verbal learning.* New York: Grune & Stratton.

Bruner, J., Goodnow, J. J., & Austin, G. A. (1967). *A study of thinking.* New York: Science Editions.

Charlesworth, R., & Lind, K. K. (2003). *Math and science for young children* (4th ed.). Albany, NY: Delmar.

Joyce, B., Weil, M., & Calhoun, E. (2000). *Models of teaching.* Needham Heights, MA: Allyn and Bacon.

Karplus, R., & Thier, H. D. (1967). *A new look at elementary school science—Science Curriculum Improvement Study.* Chicago: Rand McNally.

Martin, R., Sexton, C., & Gerlovich, J. (2001). *Teaching science for all children* (3rd ed.). Needham Heights, MA: Allyn and Bacon.

Novak, J., & Gowen, D. B. (1984). *Learning how to learn.* Boston: Cambridge University Press.

Piaget, J. (1954). *The construction of reality in the child.* (Margaret Cook, trans.). New York: Basic Books.

Vygotsky, L. S. (1987). Thinking and speech. In R. W. Reiber & A. S. Carton (Eds.), N. Minick (trans.), *The collected works of L. S. Vygotsky: Vol. 1. Problems of general psychology* (pp. 39-243). New York: Plenum.

Chapter One
Children's Acquisition of Mathematics and Science Knowledge

As they explore their environments, children actively engage in acquiring basic math and science concepts. "Science and mathematics fit together in a natural and very functional way. Mathematics is an essential component of communication for scientists. It also provides an effective way for children to process and share their discoveries" (Winnett, Rockwell, Sherwood, & Williams, 1996, p. 7).

Children's construction of knowledge builds on both content and process skills. Science is both knowledge (content) and ways of finding information and answers (process). As they use process skills, children develop concepts. For math and science, concepts and process skills are interrelated and interdependent. Charlesworth and Lind (2003) identify the basic math concepts of comparing, classifying, and measuring as the basic process skills of science. Science process skills of observing, communicating, inferring, and controlling variables are important skills for solving problems in both mathematics and science. In addition, such basic math concepts as comparing, sorting, counting, estimating, measuring, and graphing are used when solving science problems. As children deal with new situations, they begin to apply these basic concepts and process skills, such as observing, counting, recording, and organizing as they collect data.

What Is Early Childhood Mathematics?

Mathematics is a part of children's daily lives; therefore, they should be provided with a strong foundation in it. According to the National Council of Teachers of Mathematics (NCTM, 2000), "Mathematics learning builds on the curiosity and enthusiasm of children and grows naturally from their experiences" (p. 73). From birth to grade 2, cognitive growth in children is strong. They learn by thinking, doing, collaborating, sharing, and communicating about their experiences. Educators need to have high expectations of children and be aware of the many ways they learn mathematics (NCTM, 2000).

Children gain an informal understanding of many mathematics concepts before they enter school. Once children reach school age, these mathematical concepts become solidified as children continue to explore their world through both formal and informal (play) experiences. As their daily routines follow certain sequences and patterns, children experience the concept of time. They experience order as they discover and practice patterning. Children experience distance as they travel to and from school and move about their classrooms and school buildings. They experience sorting, comparing, creating sets, matching, one-to-one correspondence, classifying, counting, problem-solving, and graphing as they determine how many students are present each day; how many napkins are needed for a group of students; the number of girls versus boys in their classrooms; the number of sunny, rainy, and cloudy days each week/month; and how tall their new plants have grown. Children experience geometry and improve their spatial relationships as they explore with blocks and puzzles, and as they arrange their school supplies or classroom materials in their limited desk space, or on shelves or in boxes. Children experience measurement (length, volume, weight) when manipulating sand, water, and clay. They learn about money values when purchasing milk, lunch, and/or school supplies. Children also work with calculators to explore number sense and patterns. Clearly, children's lives involve the use of numerous math concepts on a daily basis. Nevertheless, they need adult support in order to formalize critical early childhood math concepts into the foundation for more abstract math concepts that they will encounter as they continue their education.

Careers rely upon a foundation of mathematical knowledge; therefore, everyone needs a solid understanding of mathematics. Our world continues to progress and change. In order to improve the futures of our children, we must give them the opportunity and support necessary to learn and understand mathematics. Doors will open for those with mathematical competence (NCTM, 2000).

The NCTM Publication *Principles and Standards for School Mathematics* (2000) addresses six overarching themes: equity, curriculum, teaching, learning, assessment, and technology. The Teaching Principle emphasizes that "effective mathematics teaching requires understanding what students know and need to learn and then challenging and supporting them to do it well" (p. 16). The Learning Principle emphasizes that "students must learn mathematics with understanding, actively building new knowledge from experience and prior knowledge" (p. 20). The teaching/learning models presented in the following chapters support these NCTM principles.

In addition to promoting principles and standards for mathematics, the NCTM *Standards* (2000) also makes recommendations about how educators can provide early childhood-age children with a solid cognitive and affective foundation in mathematics. A summary of the NCTM *Standards* for grades PK-12, and expectations for early childhood-age children (PK-Grade 2) (pp. 108-141), follows:

- **Number and Operations**
  - Understand numbers, ways of representing numbers, relationships among number, and number systems (e.g., count with understanding, recognize "how many," initial understanding of base 10 number system, ordinal and cardinal numbers, connect number words and numerals, understand and represent fractions 1/2, 1/3, and 1/4)
  - Understand meanings of operations and how they relate to one another (e.g., understand the meanings of "addition" and "subtraction," understand situations that entail multiplication and division)
  - Compute fluently and make reasonable estimates (e.g., use a variety of methods and tools to compute—objects, estimation, mental computation, paper and pencil, and calculators)
- **Algebra**
  - Understand patterns, relations, and functions (e.g., sort objects with similar attributes, extend patterns, analyze repeating patterns)
  - Represent and analyze mathematical situations and structures using algebraic symbols (e.g., illustrate commutative property, develop understanding of symbols)
  - Use mathematical models to represent and understand quantitative relationships (e.g., make models to represent and solve problems)
  - Analyze change in various contexts (e.g., describe qualitative and quantitative change)

- **Geometry**
  - Analyze characteristics and properties of two- and three-dimensional geometric shapes and develop mathematical arguments about geometric relationships (e.g., recognize and sort shapes, describe attributes and parts of shapes, form new shapes from other shapes)
  - Specify locations and describe spatial relationships using coordinate geometry and other representational systems (e.g., develop spatial understandings such as direction, distance, location, and representation)
  - Apply transformations and use symmetry to analyze mathematical situations (e.g., recognize and apply slides, flips, turns, and items with symmetry)
  - Use visualization, spatial reasoning, and geometric modeling to solve problems (e.g., create mental images of geometric shapes, recognize and represent shapes from different perspectives, relate geometric ideas to number and measurement, recognize geometric shapes in the environment)
- **Measurement**
  - Understand measurable attributes of objects and the units, systems, and processes of measurement (e.g., recognize, compare, and order objects according to length, volume, weight, area, and time; measure using nonstandard and standard units of measure; select an appropriate unit and tool for measurement)
  - Apply appropriate techniques, tools, and formulas to determine measurements (e.g., measure with multiple copies of unit, use repetition of a single unit to measure, use tools to measure, develop common references for measures)
- **Data Analysis and Probability**
  - Formulate questions that can be addressed with data, and collect, organize, and display relevant data to answer them (e.g., pose questions and gather data; sort, classify, and organize data about objects; represent data with objects, pictures, and graphs)
  - Select and use appropriate statistical methods to analyze data (e.g., describe data to determine what the data show)
  - Develop and evaluate inferences and predictions that are based on data (e.g., discuss informally at this level)
  - Understand and apply basic concepts of probability (e.g., answer questions informally as to the likelihood of events)
- **Problem Solving**
  - Build new mathematical knowledge through problem solving (e.g., students develop strategies to address their questions and problems)
  - Solve problems that arise in mathematics and other contexts (e.g., use interesting and challenging problems from everyday routines, new content, and children's literature)
  - Apply and adapt a variety of appropriate strategies to solve problems (e.g., students develop and implement strategies such as using manipulatives, acting it out, drawing a picture, making a list)
  - Monitor and reflect on the process of mathematical problem solving (e.g., students confirm mathematical concepts by reflecting on, explaining, and justifying their answers)
- **Reasoning and Proof**
  - Recognize reasoning and proof as fundamental aspects of mathematics (e.g., students recognize that all mathematics can and should be understood)
  - Make and investigate mathematical conjectures (e.g., students address questions to build on what they already know)
  - Develop and evaluate mathematical arguments and proofs (e.g., students consider other ideas and make arguments to support their statements)
  - Select and use various types of reasoning and methods of proof (e.g., students determine proofs through examples and nonexamples—form rules based on examples)

- **Communication**
  - Organize and consolidate mathematical thinking through communication (e.g., students explain their thinking)
  - Communicate mathematical thinking coherently and clearly to peers, teachers, and others (e.g., students talk and work with, and listen to, their peers—in pairs and small groups)
  - Analyze and evaluate the mathematical thinking and strategies of others (e.g., students evaluate mathematical thinking for understanding both content and process; students listen attentively to each other, question others' strategies and results, and ask for clarification)
  - Use the language of mathematics to express mathematical ideas precisely (e.g., students use appropriate conventional mathematical vocabulary)
- **Connections**
  - Recognize and use connections among mathematical ideas (e.g., students connect numeral names and vocabulary to concrete objects)
  - Understand how mathematical ideas interconnect and build on one another to produce a coherent whole (e.g., data collection can lead to sorting, graphing, and comparison of objects)
  - Recognize and apply mathematics in other contexts (e.g., recognize music patterns in songs and symmetry in art, collect and classify leaves on a nature walk, read children's literature about various topics with math connections)
- **Representation**
  - Create and use representations to organize, record, and communicate mathematical ideas (e.g., objects, natural language, drawings, diagrams, gestures, and symbols)
  - Select, apply, and translate among mathematical representations to solve problems (e.g., a relevant story problem solved by drawing an array also could be translated into repeated addition or a multiplication equation)
  - Use representation to model and interpret physical, social, and mathematical phenomena (e.g., a relevant story problem that involves addition with regrouping could be represented with the use of base ten blocks and/or in symbolic form)

Research on cognition reveals that children have an innate ability to learn math. Children come to school with a natural curiosity about quantitative events as well as some informal problem-solving skills (Ginsburg & Baron, 1993). Educators must effectively direct and build upon these informal skills in order to provide children with opportunities to construct meaningful understandings of mathematics.

What Is Early Childhood Science?

We are living in an era often labeled as the "Information Age." With technological advancements, including enhanced communications, science information continues to become more available to the public. With these advanced means to acquire knowledge, more scientific truths are uncovered, revealing the tentative nature of science.

The challenges that we face today may differ greatly from those our children will face when they reach adulthood. Recognizing that science is not a static body of knowledge, educators must prepare their students to question, think critically, problem solve, and make well-informed decisions. Along with the development of thinking skills, educators also must foster students' scientific attitudes such as curiosity, open-mindedness, a positive approach to failure, and a positive attitude toward change (always a necessity when considering the nature of science). In order to do so, scientific inquiry, through the use of process skills, must be taught beginning at the early childhood level.

Process skills are the "doing" parts of science that early childhood educators must promote if children are to have opportunities to discover and reinvent science knowledge. The process skills most appropriate for early childhood-age children are observation, comparison, measurement, classification, and communication (Charlesworth & Lind, 2003). These skills are developed through hands-on (concrete) experiences that encourage children to question and investigate phenomena. Students will be able to apply these skills to perform more advanced process skills as they gather, organize, and record data; infer relationships; predict outcomes; hypothesize; and identify and control variables. Some early childhood-age children are ready to explore such higher level process skills. Teachers should challenge children to stretch their abilities whenever appropriate.

Hands-on, "minds-on" science teaching methods at the early childhood level promote communication through talking, drawing, drama, puppetry, and writing. With these additional opportunities to communicate, children's development of language and reading skills improve as they build their science knowledge and become more scientifically literate.

The report *Science for All Americans* (Rutherford & Ahlgren, 1990) defines a scientifically literate person as

> one who is aware that science and technology are human enterprises with strengths and limitations; understands key concepts and principles of science; is familiar with the natural world and recognizes both its diversity and unity; and uses scientific knowledge and scientific ways of thinking for individual and social purposes. (Rutherford & Ahlgren, p. ix)

Educators need to provide children with opportunities to experience this developmental process when they are very young in order to ensure that the process continues throughout life. Science is for *all* students and must be designed to actively engage learners of all ages to help them achieve success, now and in the future.

Common areas of science for early childhood-age children to experience through process skills are life science, physical science, earth science, and health science. In life science study, children have opportunities to explore plants, animals, life cycles, and ecology. Physical science includes exploration of force, motion, energy, and machines; children enjoy exploring with levers, magnets, and changing matter. Earth science lends itself to exciting discovery opportunities through investigations about weather, air, land, water, rocks, and the solar system. In health science, children study the relationship of body parts and their functions, body systems, food, and nutrition.

Children are naturally intuitive. As a result of his research with children, Piaget found that by approximately age 4, children can fuse together knowledge and understand that one fact explains another (Gorman, 1972). Children have insights, ask questions, solve problems, and try out new ideas. Just as scientists use their minds and imaginations to create theories and hypotheses to explain their observations (Howe & Jones, 1993), children use their minds and senses to see patterns around them. They are natural scientists as they eagerly disassemble, poke, peel, and put their hands and feet into various elements. They are filled with wonder and curiosity. According to Charlesworth and Lind (2003),

> curiosity is thought to be one of the most valuable attitudes that can be possessed by anyone. It takes a curious individual to look at something from a new perspective, question something long believed to be true, or look more carefully at an exception to the rule. This approach that is basic to science is natural to children. (p. 69)

Rutherford and Ahlgren (1990) recommend that science be taught by reducing the amount of material covered and devoting more time to developing thinking skills. They also promote making connections among science, mathematics, technology, and other curriculum areas; teaching with questions rather than with answers in order to actively engage students in designing and implementing investigations; encouraging students' creativity and curiosity; and focusing on the needs of all children. Howe and Jones (1993) recommend that, when teaching children science, teachers also focus on helping children develop and maintain a curiosity about the world around them, relate what they learn in school to their own lives, and enjoy science and develop positive attitudes about learning.

Early childhood teachers should create a learning environment that supports children's curiosity and questions. When teachers help children learn about their world through hands-on, minds-on activities, they reinforce children's natural interests and curiosity while strengthening their reasoning abilities. Early childhood teachers should present science activities in ways that are natural and interesting for children. In addition, early childhood teachers need to believe in their own ability to learn new things and be open to creative methods for teaching science. It is the teacher's responsibility to provide appropriate learning situations, and then teach children how to function effectively within their boundaries (Howe & Jones, 1993).

According to Chaillé and Britain (1997), children need to experience a physical and social environment that is conducive to their work and that facilitates theory building. "Science is doing and thinking and making the two come together" (Kellough, Carin, Seefeldt, Barbour, & Souviney, 1996, p. 411). Children are mentally and physically active as they continually engage in the process of theory building and pursue their natural desire to make sense of their experiences. As children of all ages develop theories about the world, they need to be willing to build on, or often relinquish, prior ideas in order to accommodate new knowledge. During the theory building process, children's preconceptions and misconceptions are often challenged through conflict, contradiction, and trial and error. It is important for adults to remember that while some children's misconceptions can be corrected during this process, some will require more advanced thinking skills.

When designing educational science experiences that focus on learning in a social context, teachers need to remember that that there may be more activity and noise in the classroom than when knowledge is simply transmitted. Although a classroom of science learners need not sound like a circus, neither should silence be expected. The ultimate goal is for children to construct their own science knowledge as they become productive, scientifically literate citizens.

Acquiring Mathematics and Science Knowledge

One key element for children in acquiring mathematics and science knowledge is active, creative, intellectual engagement. It is vital that teachers apply a variety of teaching methods to encourage diverse populations of students to think and reason. According to the National Association for the Education of Young Children (NAEYC) guidelines on developmentally appropriate practice (Bredekamp & Copple, 1997), good teachers will "incorporate a wide variety of experiences, materials and equipment, and teaching strategies in constructing curriculum to accommodate a broad range of children's individual differences in prior experiences, maturation rates, styles of learning, needs, and interests" (p. 18). Teachers must be flexible and willing to try a variety of teaching techniques if they are to achieve meaningful learning.

The teaching/learning models described in this book provide students with opportunities to creatively construct math and science concepts while incorporating critical thinking, problem-solving, and decision-making skills, all of which are interrelated. The teaching models presented also provide opportunities for students to explore concepts—physically and mentally—use materials, and share ideas as they problem solve. In addition to learning math and science concepts, children's reasoning abilities will be enhanced through creative thought.

Before discussing these engaging teaching models, it is important to understand how children acquire mathematics and science knowledge. Considering that both the cognitive and affective domains can strongly influence how a learner experiences math and science, they are both addressed. The underlying models of cognition that are relevant when presenting mathematics and science to children also are discussed.

Cognitive Domain Theories Relevant to Mathematics and Science Learning

Jean Piaget (1898-1980)

Piaget's Knowledge Construction. Swiss psychologist Jean Piaget continues to exert a tremendous influence on educators in the area of children's thinking. Flavell (1963) described Piaget as "primarily interested in the theoretical and experimental investigation of the qualitative development of intellectual structures" (p. 15). Piaget's goal for education was to help people develop the ability to do new things and become creative inventors and discoverers (Winnett, Rockwell, Sherwood, & Williams, 1996, p. 2).

Piaget's focus was on "how" children learn. Piaget studied the thought processes of children for over 50 years to determine what goes on in their minds as they attempt to make sense of their world (Gorman, 1972). According to Piaget, children are naturally curious and intrinsically motivated to learn from their environment. Children learn because their minds are designed to learn (Ginsburg & Baron, 1993, p. 5).

Through his research with children, Piaget developed the theory that intelligence is based on logical and mathematical abilities. He theorized that children gain knowledge as they experience the world and adapt to it (Morrison, 2000). Piaget's adaptive process incorporates two interrelated processes—assimilation and accommodation. Piaget theorized that children construct knowledge by fitting (assimilating) the new information being taught into their existing models of reality (schemes) by relating new experiences to prior experiences. The term "constructivism" refers to the active building of meaningful knowledge from one's own experiences. This process requires both action and reflection on the part of the learner. Constructivists view teachers as guides who provide the setting, organize the challenges, and guide the conversation and thought processes around children's views (Shaw & Blake, 1998). Constructivist classrooms focus on the learner, not the teacher. The Association for Childhood Education International, National Council of Teachers of Mathematics (NCTM), National Science Teachers Association (NSTA), and National Association for the Education of Young Children (NAEYC) all promote such learner-centered instruction for children.

What children learn depends on what they already know (Howe & Jones, 1993). When children experience "not knowing" or "not understanding," they are set off-balance and experience a state that Piaget described as disequilibrium. During disequilibrium, learners experience a desire or need to "know" in order to return to a state of mental balance (equilibration). By creating disequilibrium within learners, teachers provide opportunities for students to modify existing schemes by including new knowledge and perspectives (accommodation). Opportunities for learning occur when children experience disequilibrium, and then seek a balance. Piaget challenged teachers to provide children with many opportunities to experience disequilibrium by presenting meaningful problems for investigation that would support spontaneous research and lead to practical knowledge (Slavin, 1991).

Piaget's Stages of Mental Development. Piaget emphasized that it takes many years for children to develop logical thinking. Piaget identified four stages of mental growth in children: sensorimotor (birth-age 2), preoperational (ages 2-7), concrete operations (ages 7-11), and formal operations (ages 11-adult). Most early childhood learners (ages 3-8) are either at the preoperational level or the concrete operational level; children often fluctuate between levels.

Piaget theorized that children at the *sensorimotor* stage experience learning through their senses and motor reflexes as they perform actions such as sucking, grasping, and other gross-body activities. As children begin to crawl and walk, they discover on their own and learn to think for themselves. In addition, children at the sensorimotor level are very egocentric; they believe they cause events.

Piaget believed that concepts of mathematics begin to develop when children grasp objects of different sizes and touch and move differently shaped objects. During infancy, children demonstrate the beginning signs of classification skills by sorting objects into different categories (Kellough et al., 1996). When children recognize that a dog differs from a cat or other animals, they are classifying. They also are very dependent on concrete representations for information, rather than relying on symbols such as numerals and words.

During the *preoperational* period, knowledge grows at a rapid pace as children develop basic concepts and skills that later lead to mathematics and science knowledge. Children at this level continue to be egocentric, believing that everyone acts and thinks the same way that they do for the same reasons. As a result, it is often difficult for preoperational children to be empathic or sympathetic (Morrison, 2000). One way to assist children's growth in this area is to provide them with opportunities to listen to others and paraphrase what others say.

Piaget discovered that children have their own reasons for thinking about things the way they do, and they are naturally resistant to change. As a result, trying to convince preoperational children to change their minds can be difficult (Howe & Jones, 1993). Furthermore, this is the stage when children start asking "why." Children at this level experience an increase in language development and are less dependent on sensorimotor actions. They should be encouraged to use all of their senses and find words to describe their perceptions in support of their emerging language and literacy skills (Winnett et al., 1996).

Children at the preoperational level can successfully experience one-to-one correspondence (matching a shoe for each sock, a napkin for each plate, etc.) and classify objects according to specific characteristics, such as color, texture, shape, size, etc. They also experience "centration"; in other words, children at this stage tend to focus or center their attention on how they perceive things, usually focusing on the first characteristic or attribute that they perceive, such as color, size, shape, etc. "Adults can verbalize their own thought processes for sorting and classifying in daily activities, thus helping children focus on attributes and how objects in one set share characteristics of another set" (Shaw & Blake, 1998, p. 28). Between the ages of 5 and 7, children experience a transition to the concrete operational stage.

At the *concrete operational* stage, children can mentally and physically carry out their thoughts through the use of symbols (words and numerals). Considering that "mathematical knowledge is a relationship constructed by the mind" (Kellough et al., 1996, p. 192), students need to construct concepts through their own firsthand experiences, interactions with others, use of language, and reflection. Students can reinvent knowledge by acting on objects (manipulatives) that they can personally handle and explore. Children who have a wide range of interests and abilities can use blocks, chips, stones, sticks, sand, clay, play dough, mud, and water for exploration. A variety of commercial educational manipulatives are available for math and science exploration, such as Unifix cubes, base ten blocks, fraction bars, magnifying lens, magnets, etc. For math, many children still prefer to use the most convenient manipulative—their fingers.

Objectivity increases in children during the concrete operational stage, allowing them to experience "if-then" reasoning; such reasoning is often used in critical thinking models and science investigations (Shaw & Blake, 1998). Children in the concrete operational stage can conserve; that is, they can determine that the quantity or quality of an object does not change simply because its physical appearance or arrangement changes. Before children can learn to conserve, they engage in counting, one-to-one correspondence, shape and space relationships, comparing, seriation (ordering items), and classification (grouping). Furthermore, children at the concrete operational level can mentally reverse their calculations and actions (reversibility). They can mentally undo what they put together, such as addition/subtraction and multiplication/division. Providing children with concrete experiences on their developmental level will better facilitate their accurate formation of mathematics and science concepts. By incorporating manipulatives into math and science curricula, students who learn by doing and being physically involved (kinesthetic modality) and by touching objects (tactile modality) will more likely have their learning styles met. In general, "children prefer to learn by touching objects, by feeling shapes and textures, by interacting with each other, and by moving things around" (Kellough et al., 1996, p. 21). The profound, insightful Chinese proverb sums it up nicely: "I hear, I forget. I see, I remember. I do, I understand."

Piaget theorized that children at the *formal operations* stage can deal with increasingly complex problems by reasoning scientifically and logically as they continue to develop new schemes (Morrison, 2000). Formal operational thinkers are less dependent on concrete objects to solve problems; they can internally visualize the manipulation of mathematical symbols (Shaw & Blake, 1998).

Bruner's Representation of Knowledge
According to Jerome Bruner (1966), the course of intellectual development for children or adults learning new material progresses through three stages: enactive (concrete—actions on objects), iconic (pictorial—visuals/images), and symbolic (abstractions—words, numerals). Unlike Piaget's stages of cognitive development, Bruner's stages may be applicable to all children and are not limited to specific age ranges. Bruner focused on how children think, as well as how children learn and how they can best be helped to learn (Howe & Jones, 1993, p. 32).

By the age of 3 or 4, most children demonstrate an interest in mathematics (Maxim, 1989). At the *concrete* level, Bruner's first stage of learning, teachers need to provide children with numerous opportunities to act on objects (manipulatives). Providing appropriate hands-on, minds-on, relevant learning experiences in both math and science can fuel this learning level. It is extremely important to remember, however, that manipulatives by themselves do not teach. Manipulatives must suit the developmental level of the child and close the gap between informal and formal school mathematics and science (Smith, 2001). Connections between informal and formal math and science understandings are critical. Teachers can facilitate these connections by carefully observing children while they work with manipulatives and other materials, and by encouraging them to share their understandings through various forms of communication such as conversation, writing, or drawing pictures.

At the transitional or *pictorial* level, described by Bruner as the second stage of knowledge representation, learners can express their understandings through conversation or by creating a mental image or picture of their concrete understanding. Children should be provided with opportunities to draw a picture/image of what they previously acted on (manipulatives), and then be asked to explain their drawings. Educators should listen carefully to children's explanations and ask them to justify their responses, whether or not they are correct. Simply allowing children to participate at the concrete level by acting on objects is not enough. "Careful and deliberate connections between manipulatives and the underlying concepts they are designed to illustrate are crucial to the construction of useful mathematical understanding" (Ginsburg & Baron, 1993, p. 15).

Children need to be provided with many opportunities to share their understandings at the transitional/pictorial level. Real objects, used in conjunction with pictorial representations, would assist learners during this transition. Teachers can use commercially made pictorial materials, or they can create their own from magazine pages, picture books, etc. Teachers need to be aware of children's understandings about each concept or topic in order to provide appropriate guidance that will enable learners to construct knowledge. When children have experienced success at the transitional level, they are ready to move on to the third level, which Bruner labeled the abstract or symbolic level.

At the abstract/symbolic level, children are introduced to symbols that represent their mental understandings, such as numerals and mathematical signs (+, -, x, =, etc.) for number concepts and letters for words. Both the concrete and pictorial stages must be nurtured before moving to the symbolic stage. Unfortunately, many teachers begin at the symbolic stage and attempt to transmit knowledge of letters, numerals, and other symbols that can create confusion within children's minds. Formal knowledge should not simply be imposed on children.

According to Smith (2001), "Bruner's three modes are found in today's math instruction: physically doing math with manipulatives; doing mental math by thinking in terms of memories of visual, auditory or kinesthetic clues; and finally being able to use number symbols with meaning" (p. 15). By allowing children to progress through all three stages of learning, teachers provide children with opportunities to take ownership of their knowledge by re-creating, reinventing, reconstructing, and redefining concepts on their own. At this point, children should have a personal understanding of the concepts that accompany symbols.

Both Bruner and Piaget believed that instruction should include a variety of appropriate materials in order for students to represent new knowledge through actions, drawings, or words (Howe & Jones, 1993). Both theorists advocated for giving children opportunities to discover concepts for themselves. Bruner promotes the various benefits of discovery learning. First, discovery learning provides opportunities for children to increase their intellectual potency by learning how to learn. Children develop skills in problem solving, which enable them to apply what they have learned to new situations. Second, discovery learning focuses on satisfying oneself rather than others. They shift from seeking extrinsic rewards to seeking intrinsic rewards. Third, knowledge that results from discovery learning is more easily recalled and remembered (Kellough et al., 1996).

Gardner's Multiple Intelligences Theory
Howard Gardner's theory of *multiple intelligences* (1999) considers intelligence to be more complex than mere capacity for storing, retrieving, and processing information. Gardner proposes that humans can be intelligent in many ways, and teachers must be attentive to the various ways that children demonstrate ability. Gardner identifies the following eight intelligences: Verbal/linguistic—intelligence with words and language; Logical/mathematical—intelligence with sequential thinking and numerical reasoning; Bodily/kinesthetic—knowledge of how to use one's body and its movements; Visual/spatial—intelligence working with images and the ability to see their interrelationships; Musical/rhythmic—intelligence with sound and musical patterns; Interpersonal—intelligence in dealing with human interactions; Intrapersonal—intelligence about one's self that leads to growth; and Environmentalist/naturalist—intelligence of dealing with the physical environment (Isenberg & Jalongo, 2001).

Patterns for learning and areas of strength emerge early in life. Some children will demonstrate intelligence in all eight areas, while others will be stronger in some than in others. Early childhood educators must be aware of their students' preferred learning styles, interests, habits, likes and dislikes, and lifestyles.

When teaching on the early childhood level, most teachers can easily incorporate Gardner's eight intelligences across the mathematics and science curriculum. Linguistic intelligence can be demonstrated through journaling, speaking, and reading about mathematics and science discoveries and/or problem solving. Logical-mathematical intelligence can be demonstrated with problem-solving, manipulating numbers, exploring relationships, and designing and conducting science experiments. Visual/spatial intelligence is used when drawing, building, designing, and creating math and science projects or creating block designs, and when creating patterns with manipulatives. Musical/rhythmic intelligence can be used to discover patterns in math activities and in nature, and from experiences with musical finger plays and poems about math and science topics. Bodily/kinesthetic intelligence can be used for movement with counting activities and for demonstrating the ability to coordinate movements when acting out knowledge acquired from scientific discoveries. Interpersonal intelligence is used when working and socializing in various cooperative learning groups that promote math and/or science discovery learning. Intrapersonal intelligence emerges when children reflect on their learning experiences and their ability to learn. Naturalist intelligence is demonstrated when children's natural curiosity is addressed by investigating life science, and by discovering and comparing features of the world.

Early Childhood Mathematics and Science Learning

Quantitative events are commonly found in children's environments across widely diverse cultures. According to Ginsburg and Baron (1993), preschool age children display a natural curiosity concerning quantitative events. They are interested in mathematics and numbers. "Children's natural curiosity about their world leads them to explore concepts of mathematics and the science of pattern and order" (Kellough et al., 1996, p. 189). Children spontaneously construct informal mathematics on a daily basis as they count, identify numbers, make comparisons, classify, and put things in order. Statements such as, "He has more than me," "She took one of my cookies," "Dad can reach higher than me," "The round pieces are mine," and "He gets to go first because he's the biggest" demonstrate informal understanding of mathematics concepts. Informal mathematics is made up of perceptions and invented strategies that deal with everyday quantifications (Kellough et al., 1996).

Children's informal mathematical conceptions provide the foundation for written mathematics learned in school. It is important for early childhood teachers to help children form meaningful connections between informal and formal mathematics. Formal mathematics includes rules, methods, and procedures for solving math problems, and is usually acquired through adult interaction (Kellough et al., 1996).

The same progression from informal to formal understanding is very evident with children's science knowledge. Children arrive at school with a wide variety of preconceptions and misconceptions about their environment. Early childhood educators need to identify children's understandings and design learning experiences that will provide them with numerous opportunities to discover, re-create, and reinvent concepts on their own, while correcting any misconceptions. According to Winnett, Rockwell, Sherwood, and Williams (1996), "misconceptions are corrected through additional experiences rather than through correction of their understanding of words" (p. 7).

Problem Solving

Problem solving is the central focus of the mathematics curriculum and the hands-on, minds-on component of science explorations and investigations. Opportunities for solving problems are abundant in early childhood classrooms, providing students with excellent opportunities to practice creative and critical thinking, as well as decision-making, skills. These skills will benefit them throughout their lives. Early childhood educators must provide children with problem-solving opportunities that instill a desire to explore, and they must expose children to productive strategies and skills that will help them develop positive attitudes toward problem solving.

Each teaching model in this book requires students to demonstrate various problem-solving skills, methods, and strategies such as persistence, tolerance of ambiguity, use of related knowledge, use of logical reasoning, pattern recognition, trial and error, as well as dealing with data, planning a solution, solving a challenge, analyzing and evaluating solutions, and working cooperatively. Incorporating real-life situations into the curriculum will give children a chance to sense problems to be solved. In order to prepare children to address the challenges that they will face throughout their lives, relevant problem-solving opportunities that stimulate creative and critical thinking must be threaded throughout the curriculum, beginning at the early childhood level.

Problem solving is a way of life. Even the youngest of children face problems daily; therefore, teaching problem solving must begin when children first enter school, and continue throughout the entire school experience. Since the emphasis is on the process and not merely the product, problem discussion, potential solutions, and methods of attacking problems should be considered across the curriculum at all times.

Problem solving can be an emotional process. Initial feelings of enthusiasm can change to feelings of frustration if each attempt meets with failure. Therefore, teachers must establish an atmosphere of success by introducing relatively simple, yet relevant, problems for children to solve. If children experience success with problem solving, they are more likely to explore solutions to new problems. Throughout the problem-solving process, early childhood educators should address the importance of developing children's confidence, willingness, and persistence. Children should be encouraged to take charge of their thinking and share responses courageously. It is important to note that success with problem solving must be earned, not simply given.

Imagination and creative thought should be encouraged throughout the problem-solving process. Problems should be created that require action, in order to encourage children's active participation. If children are directly involved in the problem-solving process, rather than mere spectators, they are more likely to learn and retain problem-solving skills.

Early childhood educators themselves must demonstrate interest, energy, and enthusiasm in order to develop children's problem-solving abilities. A positive attitude toward the problem-solving process is necessary in order to reach success. Educators must be confident in class and exhibit the same enthusiasm for the problem-solving process that they wish to instill in their children.

Careful preparation for problem solving is also necessary; educators must be aware of the varied problem-solving opportunities that present themselves in everyday classroom situations. In addition, teachers must present problems that are interesting and appealing to children. Real-life situations are very attractive to children and can be found by researching local newspapers or magazines pertinent to children's interests. Cross-curricular activities can provide challenging and enriching problem-solving opportunities. Additionally, children themselves are a source for creative, practical problems in the classroom setting.

Relevant experiences involve many opportunities for team problem solving and group brainstorming. Children should be encouraged to work in pairs or in small groups while contributing to discussions. In one approach, children in a pair could think aloud, voicing their thoughts as they attempt to solve problems. Early childhood educators should encourage children to share as many ideas for solving each problem as possible. The group method is a productive way to encourage children to respect one another's abilities and ideas while searching for a variety of ways to solve problems. Furthermore, reflection, applied immediately after a problem has been solved, is a valuable tool for children in remembering newly discovered knowledge that can be applied later to similar situations.

The early childhood curriculum should incorporate learning opportunities that will better prepare children to encounter success in the real world. Although we cannot predict the role that technology and the challenges of the future will play in our children's lives, we can safely assume that problem-solving skills and strategies will indeed improve their basic life skills and relationships with others.

Affective Domain Issues Relevant to Learning

The role of the affective domain also can influence how children acquire mathematics and science knowledge. An individual's feelings and attitudes toward learning math and science can be a major component in knowledge acquisition.

Level of Mathematics and Science Efficacy

One element crucial to acquisition of math and science knowledge concerns adults' and children's beliefs about their ability to learn and do math and science. An individual's self-efficacy (level of confidence) with math and science determines success with a particular task. Early childhood educators need to feel confident in their own ability to do and teach math and science, because their level of efficacy can directly influence children's abilities and desires to learn. If children perceive that their teacher is not comfortable teaching math and/or science, their own level of confidence and success with the subject content may be compromised.

Both early childhood educators and parents need to believe that all children can learn and do math and science. Very little evidence points to any gender or racial differences in learning math and science at the early childhood level (Smith, 2001). Every child, regardless of race, gender, ethnicity, or language difference, should be able to participate fully in mathematics and science learning experiences. If adults have low expectations of children's abilities to learn math and science, however, then the children are more likely to have low motivation and achievement. In fact, we often hear parents' attempts to explain their children's lack of or low math and science achievement as "natural," because their own personal level of achievement and self-efficacy with math and/or science is low.

Another factor that adults need to consider when working with children is that needing more time to complete a math and/or science task is not necessarily an indicator of lesser ability. The need for additional time may simply be the result of a learning style that reflects caution, a desire to be thorough, or a great interest in the subject matter (Kellough et al., 1996).

Early childhood educators, unlike most educators at other grade levels, are expected to be capable of teaching all subject areas well. Yet, we know that many teachers have specific subject intelligences, preferences, and interests. Therefore, to improve student motivation and learning, teachers often team-teach or departmentalize certain subjects in order to provide children with the best available instruction and guidance. Teachers who did not have positive personal experiences with math and science as children, yet are determined to provide their own students with productive learning experiences, could find help from administration or peers. With appropriate guidance, even teachers who have had negative personal experiences with math and/or science can positively influence children's attitudes toward the subjects.

Teachers need to design activities that will challenge children's thinking. During these activities, they should initiate and guide classroom discourse. "Discourse refers to ways of representing, thinking, talking, agreeing, and disagreeing" (Shaw & Blake, 1998, p. 55). As the children offer ideas, teachers must encourage them to clarify, verify, and justify those ideas. All children's thinking must be valued in order for them to be interested in engaging in math and science activities.

Adult and peer feedback can strongly affect a child's school experiences. It is important to recognize that children learn from what they do, whether the immediate outcome appears to be positive or negative. Mistakes are powerful learning opportunities that can be very beneficial. Scientific breakthroughs are often the result of eager, persistent, determined individuals who have been willing to try new avenues in spite of occasional, or even frequent, "failures."

As educators, we need to reinforce that learning includes experiencing mistakes and learning from them. If we constantly focus on the idea of there being only "one correct way," we could foster anxiety and learned helplessness within our children (Renga & Dalla, 1993). If children are taught at a young age that it is okay to learn from trial and error and through their mistakes, perhaps more would look forward to participating in mathematics and science courses as they continue through middle school, high school, and college.

Children at the early childhood level are often egocentric; therefore, math and science anxiety may not be as prevalent at this age level. Nevertheless, it is still very important that early childhood teachers create a nonthreatening, positive, and supportive learning environment when teaching math and science. Teachers should provide numerous learning opportunities that allow students to discover math and science knowledge on their own and share possible solutions with their classmates. Allowing children to share their strategies for solving problems in mathematics and their investigative solutions in science not only provides children with opportunities to be heard, but also helps them realize that both they and their classmates can do math and science, even if specific solutions or strategies differ. Witnessing a variety of ways of knowing, doing, and learning in math and science helps to reassure children of their capabilities, while broadening their perspectives.

Level of Enthusiasm
Early childhood teachers' own behavior when teaching math and science is critical. First, teachers need to demonstrate enthusiasm when teaching math and science. If a teacher is excited about the subject content, that enthusiasm is usually relayed to his/her students. Second, mathematics and science learning itself should be relevant and enjoyable. Activities/lessons that encourage exploration and discovery of knowledge through the use of concrete objects provide opportunities for children to construct their own knowledge. When children reinvent concepts through discovery, they take ownership of their knowledge and normally retain it longer than through direct instruction lessons. When designing math and science lessons, teachers need to create activities that are interesting and meaningful to children, are at the appropriate level of difficulty, are understandable, make use of appropriate materials, and are assessed in line with the goals and objectives of each lesson.

A variety of discovery models are appropriate when teaching children mathematics and science concepts. The Learning Cycle Lesson (Karplus, 1967), Concept Attainment (Bruner, 1966), and Concept Mapping (Novak, 1984) formats will be discussed in this text, accompanied by thorough descriptions and examples of how teachers can incorporate these effective models into their early childhood curriculum.

Additional Learning Considerations

Student Interaction

Facilitating knowledge construction in a social context is in line with Vygotsky's social learning theory. Lev Vygotsky theorized that children's mental, language, and social development is supported and strengthened by others through social interaction. He believed that children, beginning at birth, seek out interaction with adults (Morrison, 2000).

Vygotsky studied and agreed with Piaget on most topics, but differed with him concerning the role of learning through social interaction. Vygotsky felt that learning is most effective when children work cooperatively with one another (Kellough et al., 1996). Through social interaction, children become more self-directed and gradually assume more responsibility. As children talk and communicate in a friendly, relaxed manner during class time, they construct concepts and enhance social competence (Shaw & Blake, 1998).

Vygotsky's noteworthy concept of the "zone of proximal development" is very significant for early childhood educators. This concept emphasizes the value of social interaction whereby an individual is capable of accomplishing more when placed with a more competent adult or peer, in comparison to what they can accomplish working independently (Kellough et al., 1996). Vygotsky believed that children learn best when tasks are set just beyond what they can accomplish on their own, and by receiving guidance from a competent adult or peer (Morrison, 2000). Gradually, the learner masters the task and/or concept.

Cooperative Learning

Cooperative learning fosters a strong sense of community. Many discovery teaching models implemented in mathematics and science instruction can promote content knowledge construction as well as provide opportunities for children to interact socially with their peers and adults. The key to a decision about incorporating group work into your classroom should be whether children will receive a challenging experience.

Children can work cooperatively with their classmates by acting as a partner during Think-Pair-Share, or by being a member of a small group. Cooperative learning groups provide abundant opportunities for brainstorming, discussion, clarifying, and synthesizing. In addition, through role-taking (materials manager, reporter, recorder, evaluator, etc.) and/or content specialization (jigsaw), each member has a specific responsibility and is held accountable for educating other group members concerning his/her specialized knowledge contribution. It is very important to remember that the seating arrangement of the classroom should support student interaction during cooperative work.

Another benefit of cooperative learning groups is the emphasis on positive group behavior. Children are reminded to listen to each group member's contributions and respond in an open, positive manner, and to allow all group members to share their ideas, share materials, and assist each other when needed. With cooperative learning activities, children in each group help their group members, rather than compete against them. Thus, cooperative learning models promote powerful social skills as well as an appreciation for contributing to the benefit of the whole, in terms of shared knowledge acquisition. Group project work often motivates children to share ideas, discuss various strategies, and disagree, when appropriate. Because of the many benefits of cooperative learning, both academic and social, teachers should provide opportunities for children to experience mathematics and science in cooperative learning groups.

Peer and cross-age tutors also can be used effectively when working with children in math and science. Children who need clarification on specific math and science content/concepts can benefit when working with a child who has a positive attitude toward math and/or science; has mastered the necessary content; and who is patient, understanding, and tolerant when working with classmates. In addition, the tutors benefit as they strengthen their knowledge base and their level of efficacy with specific content through teaching others (Gallenstein & Johnson, 1998).

Activities/Games/Puzzles
When promoting enthusiasm for learning math and science concepts, teachers often utilize playful activities, games, and puzzles. Constance Kamii (1986), a student of Piaget's, emphasizes that children need opportunities for problem solving through games and activities that challenge their minds. Many challenging, thought-provoking, higher level math and science games/activities are available for children via the Internet and on CD-ROMs. Commercially made, nonelectronic games and puzzles are also available.

The element of "chance" should be included in math and science games. If only those who have fully mastered math and science concepts can win at games/activities, those children still attempting to master concepts might very well be discouraged from playing. An element of chance serves not only as a motivator for participating in a game/activity, but also as a catalyst for sticking with the game, which should result in additional learning.

To design learning opportunities for children's specific needs, many teachers create their own games, activities, and puzzles that reinforce basic math/science concepts and process skills. Teacher-made games and puzzles can be made more durable through lamination or by adhering contact paper to the final product. These games/puzzles can be available in learning centers or on easily accessible shelves for children to use over and over and over again. Centers serve many purposes. They can be used for exploration and reinvention of concepts, drill and practice for concept reinforcement, or as an extension of concepts already understood.

Play
The element of play is another key component of a developmentally appropriate mathematics and science curriculum. Children should have time to explore materials and discover math and science concepts through play. "Free exploration and play are a need that must be fulfilled before children can see materials as learning resources" (Winnett et al., 1996, p. 8). The NAEYC publication on developmentally appropriate practice (Bredekamp & Copple, 1997) stresses that science explorations, math skills, and problem solving are fostered through play, projects, and situations of daily living.

A wide variety of materials should be available in early childhood classrooms for play that would contribute to children's mental growth in the area of math and science. Furthermore, children should have sufficient time to play with materials and not be rushed from one project or activity to another. Also, if possible, space should be made available so that children who require more time to complete their constructions can leave them in progress and return to them later. Early childhood teachers can photograph or videotape their children's creations for display, thus acknowledging the children's efforts as worthwhile.

Blocks, sand and water tables, containers of manipulatives, prop boxes, and interest centers can contribute to children's discoveries of math and science concepts. As children play, teachers must be available to listen to their conceptions, preconceptions, and misconceptions. Teachers can encourage children to "talk aloud" about how they arrived at their answers, and encourage them to practice listening skills as their peers explain their own solutions (Smith, 2001). Questioning and guidance by teachers during play can lead children to the next level of learning.

Sense of Competence
Another factor that may encourage children's positive attitudes toward learning math and science is to design activities that will allow them to succeed. "Educators must provide

activities that allow success and reinforce independence for children. Open-ended activities allow children to develop confidence in their abilities to analyze" (Shaw & Blake, 1998, p. 50).

Early childhood educators need to be aware of their children's learning levels in order to design activities that are developmentally appropriate and meaningful. Teachers should challenge, but not overwhelm, children, providing activities that range across ability levels, learning styles, and multiple intelligences. Activities should provide children with opportunities to extend their thinking and interact freely with their peers and adults, while allowing for the integration of other curriculum areas. These activities can be introduced to all the children in whole-class instruction, and then be placed in learning centers. Furthermore,

> materials should be accessible and the curriculum developed in such a way that children can return to or repeat experiences that they have completed some time ago. Repetition reinforces children's awareness of their own competence and the confidence that awareness brings. (Winnett et al., 1996, p. 3)

In order for children to fully benefit from activities, teachers should observe, ask questions, and make suggestions to them while they work at the various learning centers. Once children experience success with selected center activities, teachers then should enthusiastically recommend a more challenging activity.

Making Learning Meaningful

Another important factor that early childhood educators teaching mathematics and science must consider is that children need to understand and appreciate the relevance of what they are learning. Applying mathematics and science to the daily classroom environment allows children to value math and science, and to understand reasons for learning those subjects. If teachers connect math and science knowledge to real-life situations and career opportunities, children will have a greater appreciation for the content. Math and science concepts are important to daily life; therefore, early childhood teachers must reinforce the value of what is being learned by connecting it to meaningful experiences. Relevant learning opportunities are abundant in classrooms, school buildings, home settings, and local communities. One need not look far for ties to math and science. A field trip to a local grocery store or pizza parlor holds great promise for math and science learning experiences. Many children also delight in having their parents/guardians/relatives share aspects of their careers that deal with relevant, meaningful math and science content. Children can invite adults into their classrooms as guest speakers or arrange field trips to the adults' places of employment. Remember to keep the content age-appropriate.

Subject Integration

When promoting positive attitudes toward mathematics and science learning, early childhood teachers do well to take into consideration their children's interests. Most children are interested in learning about almost everything. They are interested in insects, furry animals, boats, cars, calculators, computers, the solar system, simple machines, etc. And children usually like to share what is on their minds. If teachers listen carefully to this sharing, they will easily identify themes or topics that are of interest to the children and relevant to their lives. It is highly recommended that teachers encourage children to discuss their interests and allow them to help design integrated thematic curriculum units. If children's interests are piqued, they will more likely be willing to participate in the designed learning experiences.

Teaching with integrated thematic units (subject integration) is a productive way to make learning relevant and meaningful for children. The human brain seeks meaningful connections when presented with new information; current knowledge of brain development strongly supports curriculum integration, rather than distinguishing among subject areas. This approach favors in-depth knowledge and understanding of a topic instead of quick, shallow learning (Bredekamp & Copple, 1997). "When learning is centered around a small number of core concepts, the learner can spend enough time with materials and concepts to master them" (Winnett et al., 1996, p. 2). When lessons are connected, ideas and concepts flow, rather than being taught as fragments that are in isolation from one another. "For higher levels of thinking and for learning that is most meaningful and longest lasting, research supports the use of an integrated curriculum and instructional techniques that involve the children in social, interactive learning" (Kellough et al., 1996, p. v). Although curriculum integration is strongly emphasized today, John Dewey advocated the approach as early as the turn of the 20th century (Shaw & Blake, 1998).

Connecting mathematics and science content to other subject areas, such as language arts, also can promote positive attitudes. Children's literature contains numerous relevant math concepts—the popular folk tale of *The Three Bears*, for example, touches upon one-to-one correspondence, size, volume, proximity, etc. Such authors as Marilyn Burns and Stuart Murphy have written exciting, entertaining, educational books with early childhood math connections. Numerous literature books with science connections are also available for early childhood-age children. It is exciting to observe a child's enthusiasm when the story and pictures in a trade book trigger mathematical and science conceptual understandings. Lists of books that support linking literature with math and science can be found in publications such as the following.

- Burns (1992), *Mathematics and Literature (K-3)*
- Sheffield (1995), *Math and Literature (K-3): Book 2*
- Thiessen, Matthias, and Smith (1998), *The Wonderful World of Mathematics: A Critically Annotated List of Children's Books in Mathematics*
- Welchman-Tischler (1992), *How To Use Children's Literature To Teach Mathematics*
- Whitin and Wilde (1992), *Read Any Good Math Books Lately? Children's Books for Mathematical Learning (K-6)*
- Whitin and Wilde (1995), *It's the Story That Counts: More Children's Books for Mathematical Learning (K-6)*

Providing opportunities for children to write and draw pictures about what they have learned and experienced with math and science can provide a teacher with the necessary insights on how a child thinks and feels. Children can create their own math and science books and regularly record their thoughts in journals or learning logs; they can use either picture or story form. According to Shaw and Blake (1998),

> children's emerging writing abilities are extended when adults take dictation of what the children say, or write experiences charts as a group of children discusses a situation. Using manipulatives, models, charts, graphs, and diagrams clarifies and stimulates communication in mathematics. (p. 10)

Teachers then can create additional meaningful experiences with math and science concepts.

Prop boxes and interest centers that include mathematics and science concepts are wonderful tools for allowing children to express their understandings, preconceptions, and misconceptions about math and science concepts. It is important that children be able to initiate their own experiences. They need to touch, handle, feel, move, pound, taste, see, hear, and do something in order to learn. Children also need time to reflect on such learning experiences. In recognition of this need, teachers should provide children with opportunities to organize and present their learning perceptions. Children also should have opportunities to share what they liked best about their experiences and what knowledge and skills they have learned. Through these experiences, early childhood teachers help children clarify their understandings and gain insight for designing additional experiences through which children can explore and discover accurate math and science concepts.

Feedback and Assessment
Teacher feedback also strongly influences how children view themselves as learners of mathematics and science. Early childhood teachers need to focus on what children can do correctly, and then facilitate learning experiences that will clarify any misunderstandings. Ongoing assessment is vital to effective teaching. Many teachers keep a clipboard or notebook close by to record children's progress. Early childhood educators must remain aware of what children know in order to prepare effectively for the next lesson/activity. Renga and Dalla (1993) state:

> Since teacher feedback is positively related to children's assessment of their abilities, teachers should provide feedback that focuses on what students do right, treats mistakes as a normal part of learning, shows them what both their skill and conceptual errors are, helps them understand why they made errors, provides them with the means for evaluating their own processes and solutions, gives suggestions for improvement, and makes them feel that they are competent. (p. 26)

Specific assessment methods used by teachers will affect children's attitudes when learning math and science content. Today, teachers are encouraged to apply various forms of alternative assessment techniques when evaluating children. Pencil-and-paper tests, which often create anxiety in children, are now supplemented with a variety of authentic, performance assessment methods such as oral questioning; whole-class, small-group, or individual discussions; progress portfolios; teacher-created rubrics; observation, accompanied by anecdotal records, checklists, and/or rating scales; journal writing; and teacher/student interviews. When appropriate, it is recommended that teachers inform children in advance of the criteria for assessment, whether they involve academics, social skills, or a combination of both. Additionally, rubrics designed for specific math and science assignments should be shared with children in advance if they are to understand what is expected of them. Portfolios, which could include both student- and teacher-selected artifacts, should also have a shared set of criteria so that everyone interested in reviewing a child's progress will have a firm understanding of student outcomes in relation to student expectations.

It appears that children's accomplishments in mathematics and science can be assessed more thoroughly and fairly with the use of alternative assessment methods. Furthermore, when authentic/performance based assessment techniques are utilized, children's various learning modalities and multiple intelligences are more likely to be evaluated.

Autonomy
While "doing" mathematics and science, children have opportunities to become more self-directed and less dependent on their teachers. The teacher is responsible for arranging the learning environment and acting as an orchestrator for children's construction of knowledge. As children become increasingly competent, their level of confidence and ability to construct knowledge increases. Considering that the responsibility of constructing knowledge ultimately lies with the learner, teachers can facilitate children's development of autonomous learning (learning to learn) by providing them with numerous opportunities to

construct their own knowledge.

Constance Kamii (1986) emphasizes Piaget's assertion that autonomy is the aim of education. In order for intellectual autonomy to develop, children need to feel secure in their relationships with adults, who should encourage them to be curious and share their ideas with other children. Adults should encourage children to initiate interesting problems, ideas, and questions; speak their minds with confidence; and attempt to tackle problems. Furthermore, as children interact socially with their classmates during math and science learning, they will experience different cultural values, ideas, perspectives, and solutions.

As we progress in this highly technical Information Age, children will continue to need strong foundations in mathematics and science. Technology, being the application of knowledge, is fueled by mathematics and science knowledge. Early childhood educators must employ methods for teaching math and science that will generate positive attitudes in their students. The world will rely upon open, eager minds to meet future needs in the area of mathematics and science. If educators live up to this responsibility, children's lives will be enhanced and society as a whole will benefit.

Interrelationship of Mathematics, Science, and Literacy

Many math/science concepts and process skills are acquired when children play with blocks, sand, and water, and engage in dramatic play, cooking, nature walks, etc. Children use math and science thinking skills such as comparison and classification as they discriminate among shapes, size, and amount. During the preschool years, children strengthen their language skills by expressing math concepts and contrasting characteristics such as big, small, light, heavy, square, round, long, short, etc. Literacy skills also can be reinforced through hands-on science as they observe, explore, discover, and reinvent concepts. These skills are valuable as children learn to read and write. They need the ability to discriminate among letters in the alphabet, syllables, and words.

For more advanced children, reading math and science background material and recording hypotheses and observations helps them develop reading and writing skills. Additionally, children strengthen their language communication skills when recording, discussing, and sharing their math and science discoveries through various means, including graphic aids, drama, and puppetry. Furthermore, practice with sequencing and cause-and-effect relationships discovered in math and science can aid children when predicting story outcomes and applying information to other situations (Charlesworth & Lind, 2003).

Summary

An understanding of children's developmental levels in relation to learning mathematics and science is critical for early childhood educators when designing effective learning experiences. When designing learning experiences, attention must be given to both the cognitive and the affective domains. Additional learning considerations, such as the role of student interaction, cooperative learning, use of games and puzzles, inclusion of play, awareness of a sense of competence, value of making learning meaningful, subject integration, providing accurate feedback and assessment, and fostering autonomy, must be considered in order to create productive environments that foster mathematics and science learning.

The next three chapters will build on the information presented in Chapter 1 by providing detailed explanations of specific teaching/learning models and examples appropriate for children when introducing and extending mathematics and science knowledge.

References

Bredekamp, S., & Copple, C. (Eds.). (1997). *Developmentally appropriate practice in early childhood programs* (Rev. ed.). Washington, DC: National Association for the Education of Young Children.

Bruner, J. (1966). *Toward a theory of instruction*. Cambridge, MA: Harvard University Press.

Burns, M. (1992). *Mathematics and literature (K-3)*. White Plains, NY: Cuisenaire Co.

Chaillé, C., & Britain, L. (1997). *The young child as a scientist: A constructivist approach to early childhood science education* (2nd ed.). New York: Longman.

Charlesworth, R., & Lind, K. K. (1999). *Math and science for children* (3rd ed.). Albany, NY: Delmar.

Flavell, J. H. (1963). *The developmental psychology of Jean Piaget*. Princeton, NJ: D. Van Nostrand.

Gallenstein, N. L., & Johnson, F. F. (1998). Experiences of elementary Hispanic students in predominantly Anglo environments: A phenomenological study. *GATEways to Teacher Education, XII*(1), 35-54.

Gardner, H. (1999). *Intelligence reframed: Multiple intelligences*. New York: Basic Books.

Ginsburg, H. P., & Baron, J. (1993). Cognition: Children's construction of mathematics. In R. J. Jensen (Ed.), *Research ideas for the classroom: Early childhood mathematics* (pp. 3-21). New York: Macmillan.

Gorman, R. M. (1972). *Discovering Piaget: A guide for teachers*. Columbus, OH: Merrill.

Howe, A. C., & Jones, L. (1993). *Engaging children in science*. New York: Macmillan.

Isenberg, J. P., & Jalongo, M. R. (2001). *Creative expression and play in early childhood*. Upper Saddle River, NJ: Prentice-Hall.

Kamii, C. (1986). *Cognitive learning and development. Today's kindergarten*. New York: Teachers College Press.

Karplus, R., & Thier, H. D. (1967). *A new look at elementary school science-Science Curriculum Improvement Study*. Chicago: Rand McNally.

Kellough, R. D., Carin, A. A., Seefeldt, C., Barbour, N., & Souviney, R. J. (1996). *Integrating mathematics and science for kindergarten and primary children*. Englewood Cliffs, NJ: Prentice-Hall, Inc.

Maxim, G. W. (1989). Developing preschool mathematical concepts. *Arithmetic Teacher, 37*(4), 36-41.

Morrison, G. S. (2000). *Fundamentals of early childhood education* (2nd ed.). Upper Saddle River, NJ: Prentice-Hall.

National Council of Teachers of Mathematics. (2000). *Principles and standards for school mathematics*. Reston, VA: Author.

Novak, J., & Gowen, D. B. (1984). *Learning how to learn*. Boston: Cambridge University Press.

Renga, S., & Dalla, L. (1993). Affect: A critical component of mathematical learning in early childhood. In R. J. Jensen (Ed.), *Research ideas for the classroom: Early childhood mathematics* (pp. 22-39). New York: Macmillan.

Rutherford, F. J., & Ahlgren, A. (1990). *Science for all Americans*. New York: Oxford University Press.

Shaw, J. M., & Blake, S. S. (1998). *Mathematics for children*. Upper Saddle River, NJ: Prentice-Hall.

Sheffield, S. (1995). *Math and literature (K-3): Book 2.* Sausalito, CA: Math Solutions.

Slavin, R. E. (1991). *Educational psychology* (3rd ed.). Englewood Cliffs, NJ: Prentice-Hall.

Smith, S. S. (2001). *Early childhood mathematics* (2nd ed.). Needham Heights, MA: Allyn & Bacon.

Thiessen, D., Matthias, M., & Smith, J. (1998). *The wonderful world of mathematics: A critically annotated list of children's books in mathematics.* Reston, VA: National Council of Teachers of Mathematics.

Welchman-Tischler, R. (1992). *How to use children's literature to teach mathematics.* Reston, VA: National Council of Teachers of Mathematics.

Whitin, D. J., & Wilde, S. (1992). *Read any good math books lately? Children's books for mathematical learning (K-6).* Portsmouth, NH: Heinemann.

Whitin, D. J., & Wilde, S. (1995). *It's the story that counts: More children's books for mathematical learning (K-6).* Portsmouth, NH: Heinemann.

Winnett, D. A., Rockwell, R. E., Sherwood, E. A., & Williams, R. A. (1996). *Discovery science: Explorations for the early years.* New York: Addison-Wesley.

Chapter Two
Making Learning Meaningful Through the Learning Cycle Lesson Format

We have all heard children say such things as, "Why do I have to learn this?," "I'll never use this!," or "I can't remember the rule." Children often feel compelled to make such comments when learning mathematics and/or science because they do not always understand the relevancy and significance of the concepts being presented. When math and science instruction is irrelevant to students' lives, they may find the subjects to be boring and meaningless. The teaching model presented in this chapter, called the Learning Cycle, promotes exploration and discovery, while demonstrating the relevancy of mathematics and science concepts. The learning cycle lesson format is an instructional model inspired by Piaget's constructivist learning theory (Kellough et al., 1996), which provides children with an opportunity to learn in a "manner congruent with how they learn naturally" (Charlesworth & Lind, 2003, p. 82).

The learning cycle lesson format originated in the early 1960s as the teaching model used in the Science Curriculum Improvement Study (SCIS) program. When the Soviets launched the *Sputnik* satellite in 1957, the United States government allocated considerable funding for mathematics and science education programs as part of the nation's efforts to remain competitive in the space race. The SCIS program, having the greatest overall effect on student achievement, was one of numerous innovative science programs that resulted from this funding (Martin, Sexton, & Gerlovich, 2001). The learner-centered, activity-based SCIS program also resulted in significant gains with early childhood-age children in the area of conservation of length and number (Kellough et al., 1996, p. 433).

A team of educators at the University of California, Berkeley, developed SCIS (Kellough et al., 1996). The program was directed by Robert Karplus, field tested during the 1960s, revised during the 1970s, and is still in widespread use today. The overall goal of this inductive instructional approach is to help learners form a broad conceptual framework for understanding science through exploration and discovery while taking ownership of concepts (Martin et al., 2001). The SCIS program emphasizes both process skills and content as students explore. The teacher's role in the learning cycle format is to facilitate and guide learning.

The original SCIS learning cycle model consisted of three distinct phases: exploration, invention, and application. In the early 1970s, I used this model for teaching early childhood science. I found it to be highly effective, and the children whom I worked with were always eager to "do" science. Since the format's introduction, many revisions and adaptations have been developed. The current revised SCIS 3 program utilizes the 4E learning cycle model, which includes four phases: exploration, explanation, expansion, and evaluation.

Another learning cycle version is the 5E model, which I promote when introducing preservice and inservice teachers to effective teaching methods for early childhood learners in both science and mathematics. The five phases of the 5E model are: engagement, exploration, explanation, expansion, and evaluation. This format is highly effective when included in thematic curriculum units, and particularly so when placed as introductory lessons in units. The lessons, as such, provide children with opportunities to build on prior knowledge, reinvent knowledge through discovery, and extend their knowledge through additional exploratory activities and lessons.

The *engagement* phase serves as a motivator or focus for each lesson. For example, the teacher creates the environment and demonstrates enthusiasm by asking a thought-provoking question, presenting a discrepant event/scenario, and/or sharing a story/picture/word that initiates thinking on the part of the children. Children attend as the teacher initiates discussions that pique their interest. This phase also can be used to assess children's current knowledge base. As children respond to the motivation phase, the teacher can move gradually forward into the next phase, *exploration*.

The *exploration* phase is learner-centered. The intent in this phase is for children to activate all of their senses. "Firsthand inquiry experiences like the learning cycle make use of a child's natural curiosity rather than trying to suppress it" (Charlesworth & Lind, 2003, p. 69). During the exploration phase, the teacher provides concrete experiences in order to stimulate learner disequilibrium and foster assimilation (Martin et al., 2001). The teacher provides items to observe, opportunities for active learning, and basic instructions for safety and material use, and then allows children to explore, discover, and reinvent concepts while constructing their own personal meanings of their experiences. During this phase, the teacher also focuses children's observations, helping them reach conclusions and form concepts as they describe and label their thoughts (Kellough et al., 1996).

During the *explanation* phase, the teacher uses focused questioning, asking children to share what information they have discovered and collected during the exploration phase. The teacher guides the children toward making connections to their prior knowledge by processing and mentally organizing their discovered knowledge. Children then construct the concept cooperatively and apply an appropriate label for the reinvented concept.

Although often used in the social studies curriculum, the questioning strategies promoted by Hilda Taba and her associates are appropriate for the explanation phase as a means by which children identify what they have learned as a result of experiencing an event. Taba recommends encouraging all children to participate in discussions by asking questions such as, "What did you think of when you heard/saw ____?," "Why did your group____ in that way?," "What comes to mind when you think of ____?," "What would happen if ____?," "What have you tried?," and "What results/conclusions could you draw from the investigation?" (Martorella, 1998). Questions during the explanation phase can encourage openness and challenge children's thinking.

Next, during the *expansion* phase, the teacher provides children with opportunities to apply learning to meaningful situations by expanding their newly acquired concept(s) through additional explorations and experiences. As with the *exploration* phase, this phase is also learner centered. Children apply knowledge to real-world experiences and/or career awareness, thus gaining an understanding of the relevancy that mathematics and science concepts have for them.

Science, Technology, & Society, or STS, also is emphasized during the expansion phase of the 5E model. STS refers to science as knowledge, technology as the application of knowledge, and society as the group affected by this application of knowledge. Learning becomes relevant as children increase their awareness of the effect science and technology can have on the population and the environment.

Although listed as the fifth phase, *evaluation* occurs throughout all phases of the 5E model in the form of teacher/student questions, discussions, and interviews; teacher-created observation checklists/rating scales or rubrics; student journaling, presentations, or projects; and/or paper-and-pencil tests, etc. Both informal and formal assessment can play a role in the evaluation phase.

The 5E learning cycle model has provided educators (both preservice and inservice) and children alike with opportunities to discover, reinvent, and understand early childhood mathematics and science concepts. As a result, when learning occurs, children take ownership of math and science concepts and appear to feel more comfortable with their mathematics/science abilities as they enthusiastically share their understanding of math/science concepts with their peers.

The 5E learning cycle format is highly effective with early childhood-age children. Sorting, comparing, classifying, and introductory number concepts, which are appropriate for children on the preschool and kindergarten levels, can be developed into 5E learning cycle lessons. (Refer to sample lessons found later in this chapter.) In an example of a 5E learning cycle lesson appropriate for early childhood mathematics, children are exposed to various interpretations for the basic operations of addition, subtraction, multiplication, and division. When experiencing the various phases of the 5E model, children have an opportunity to draw on their prior knowledge and relevant experiences and discover and reinvent math operation interpretations on their own, rather than simply being informed by their teacher through direct instruction. As a result of their explorations with manipulatives, children can define operational concepts and suggest appropriate labels. Often, to children's amazement, their definitions are accurate; as a result, they develop a strong sense of conceptual ownership. At this time, children have additional opportunities to expand their ideas and reinforce their conceptual understandings. Furthermore, children can share with others how they actually use and apply discovered mathematics/science concepts on a daily basis. They learn about careers that utilize discovered and defined concepts, thus gaining insight about the relevancy of these concepts.

A more specific math example for primary age children concerns rational numbers, which can be a challenging concept for both children and adults. Through the 5E model, children have an opportunity to explore and discover number relationships as they use manipulatives, constructing and defining conceptual understandings about parts of the whole in the form of fractions, decimals, and percents. This knowledge then can be expanded through additional experiences, depending on the children's age level, that provide opportunities to tie rational number concepts to real-world experiences and career opportunities such as cooking, sewing, gardening, medicine, construction, hourly wages, track and field events, measurement, probability, statistics, and graphing (Sheffield & Cruikshank, 1996).

Although the learning cycle format originally was designed to increase the effectiveness of teaching science concepts, it is an extremely effective teaching/learning model to use when teaching children both science and mathematics (Gallenstein, 1999). Children learn best when they are actively involved in the acquisition of knowledge (NCTM, 2000; National Research Council, 1996), and they develop a deeper understanding of concepts when they are encouraged to construct their own knowledge. As learners at all levels reinvent, define, and take ownership of math and science concepts through exploration and discovery utilizing the 5E learning cycle model, they should develop solid links to real-world experiences and career opportunities. In the process, their questions concerning the value of mathematics and science will be addressed.

The next section of this chapter contains a summary of the 5E learning cycle lesson format and provides various examples of specific mathematics and science learning cycle lessons. Some of the lessons are original creations by the individuals listed, while others were ideas or activities initially designed for direct instruction and adapted by preservice and inservice teachers to fit the 5E constructivist learning cycle lesson plan format.

5E Learning Cycle Lesson Format

(Adapted from Martin et al., 2001)

PHASE

ENGAGEMENT Motivation, Focus, Anticipatory Set
- Teacher asks a thought-provoking question, presents a discrepant event/scenario, and/or shares a story/picture/word to focus lesson
- Teacher assesses children's current knowledge

EXPLORATION Learner-Centered Phase
- Teacher allows the children to develop critical thinking skills by exploring materials and discovering concepts through concrete experiences, while stimulating learner disequilibrium
- Teacher provides basic instructions for material use and reviews appropriate safety factors

EXPLANATION Teacher-Directed Phase
- Through questioning, teacher asks the children to share information collected during the exploration phase
- Teacher guides children's thinking in order to construct and label concept(s) cooperatively

EXPANSION Learner-Centered Phase
- Teacher provides children with an opportunity to expand constructed concepts through additional concrete explorations
- Teacher provides basic instructions for material use and reviews appropriate safety factors
- Concept(s) applied/connected to real-world experiences; Science, Technology, & Society (STS); and/or career awareness

EVALUATION Formative and Summative Assessment
- Teacher evaluates previous four stages through alternative assessment such as questions, discussions, interviews, observations, checklists, rating scales, rubrics, journal writing, projects, presentations, quizzes, tests, etc.

SAMPLE LESSON

Lesson Title: Whose Shoes?
Concept: Sorting Objects by Attributes
Grade Level: PK-1st
Source: Amy Dannis (inservice teacher)
Materials: Children's shoes, pictures of shoes, the book *Shoes* by Elizabeth Winthrop.

5E PHASES

Engagement (Motivation):
- Walk around the room and unobtrusively look at everyone's feet. Do *not* tell the children what you are doing. Say, "Hmmmm.... I see some that are white and some are pink. There are a few brown ones also. Some of them tie, and some don't tie."
- Ask children if anyone can guess what you are talking about. If no one knows, add, "Some have laces, and some have Velcro."
- At this point, inform the children that you will count to three and they should say their response chorally when you reach "3."

Exploration (Hands-on, Minds-on Concept Exploration):
- Divide children into groups of five or six. Ask each child to remove one shoe and put it in the center of the circle.
- Walk around and say, "See what you can notice about the shoes. Find some things that are the same about your group's shoes. What are some ways that your shoes are alike? What are some ways that your shoes are different?"
- Once children have completed discovering similarities and differences on their own, ask them to sort the shoes in one way of their choice. Then, ask children to sort their shoes another way. Continue to walk around and ask the following questions: "How did you sort your shoes?," "Can you sort them another way?," "Why doesn't this shoe belong with that group?," and "Which group would my shoe belong in?"

Explanation (Concept Development):
- After children have had time to discover and sort, discuss the activity as a group. Ask each group to explain ways that they sorted their shoes.
- Next, ask children for other ways that the shoes could be sorted. Answer any questions before asking the following: "Are there things in the real world that you sort? What are they?," "Are there things in this class that we can sort?," "Why is sorting so important for us?," and "What would happen if we never sorted things in our classroom?"

Expansion (Concept Relevancy):
- Read the book *Shoes*, by Elizabeth Winthrop, aloud for the class. Introduce it by saying, "I have a special book to read to you that's all about shoes. Do you think that maybe they thought of ways to sort shoes that we did not think of? Let's find out."
- After the book is read, discuss it. Then ask the following questions: "What ways did they sort shoes that were the same as ours?," "Did they think of ways to sort that we did not think of? What were they?," and "Did we have any ideas that the author did not think of?"

- To extend this activity, the children could form a class circle and sort out all the different shoes. They could see that the more shoes there are, the more different ways they can be sorted.
- Another extension would be to put pictures of shoes in a learning center. In small groups or as individuals, children could see how many ways they could sort the pictures of shoes.

Evaluation (Formative/Summative Assessment):
- Circulate around the classroom while the groups are sorting their shoes and listen to their interactions. Complete a checklist with the following criteria (refer to checklist examples—Figures 1 & 2):

- Sorts by one attribute
- Sorts by two attributes
- Sorts by more than two attributes
- Sorts during teaching/child interview
- Cooperates with group members
- Participates in group activity
- Participates in class discussion
- Uses math concept terms appropriately.

Child's Name _____

Levels: + Accomplished   ✓ Developing   - Beginning

LEVELS

CHECKLIST CRITERIA:	Date	Date	Date	COMMENTS
Sorts by one attribute				
Sorts by two attributes				
Sorts by more than two attributes				
Sorts during teacher/child interview				
Cooperates with group members				
Participates in group activity				
Participates in class discussion				
Uses math concept terms appropriately				

ADDITIONAL COMMENTS:

Figure 1

DATE

CRITERIA

LEVELS:
+ Accomplished
√ Developing
- Beginning

Children's Names

COMMENTS

Figure 2

SAMPLE LESSON

Lesson Title: What's My Idea?
Concept: Classification
Grade Level: PK-1st
Source: Jackson et al. (1996). *Mathematics in Action*. Macmillan/McGraw-Hill, p. 161.
               Idea adapted by Erica Littman, Angela Tatum, and Sara Tulenson (preservice teachers)
Materials: Packet of manipulatives for each group that includes items such as buttons, shells, crayons, markers, pattern blocks, money, paper clips, cotton balls, etc.

5E PHASES
Engagement (Motivation):
- Divide the class into two groups of children by some characteristic (e.g., one group is wearing shorts and the other group is not wearing shorts). Ask the children to think about how they are divided. The children may discover more than one correct way. Acknowledge every response in a positive manner.
- In order to clarify any misunderstandings and stretch other children's thinking, provide each child with an opportunity to justify why he/she chose a particular characteristic as the sorting determinant.
- When the characteristic you used to sort has been identified, divide the class into two different groups, using another characteristic. Again, children will determine how they are sorted and share their reasons for their choices. At this time, assess each child's comprehension level concerning the activity, and determine if additional groupings are necessary.

Exploration (Hands-on, Minds-on Concept Exploration):
- Based on children's understandings of the engagement activity, divide them into small heterogeneous groups of two to three members.
- Provide each group with a bag of multiple manipulatives and inform children of appropriate and safe use of the objects.
- Inform the children that they each will have a chance to be the "sorter." Have the children decide who will be the first sorter in their group. Explain that the sorter will decide how to sort the objects and will tell his/her group members the characteristic he/she will use. The group members will sort the objects together according to the sorter's chosen characteristic.
- Group members then switch roles, and the new sorter chooses a different way to sort the objects. Be certain that each child has an opportunity to be the sorter at least once.

Explanation (Concept Development):
- As a whole class, and after children have sorted their materials, each child has an opportunity to share with his/her classmates how the materials were sorted and why.
- Then ask the children to share some other words that people use to mean "sorting" ("grouping," "dividing," "classifying"). Also, ask children how and why they think that sorting objects can be helpful.

Expansion (Concept Relevancy):
- Now pose questions that are relevant to children's lives, such as "Where have you seen things separated?" (clothes in a closet or in drawers, cash register, silverware, etc.); "What other places have you seen where things are separated?" (shopping malls, grocery store, house, etc.) "What in this room is separated/sorted into groups?" (art supplies, books on shelves, blocks, etc.); and "What outside is sorted?" (cars on the parking lot, equipment on the playground, sandbox materials, etc.). Allow children to look out the classroom windows or go outside as a class in order to investigate how things are separated.
- Again, ask children to explain why we use classification/sorting and if they think it makes our lives easier and how.
- The children can find another example of classification at home to share with their classmates on the next day. They may draw or write about their examples if they like.

Evaluation (Formative/Summative Assessment):
- Ask various questions and initiate discussions throughout the lesson, during both small- and large-group activities.
- Use a checklist to date and mark which children were or were not successful with the specific criteria for classification. Social behavior in groups also can be assessed and included in the checklist criteria.
- When appropriate, make comments (anecdotal records), recorded in a separate column, on the checklist concerning children's success level with classification.
- Also review for accuracy children's drawings and writing samples of classification from home.

SAMPLE LESSON

Lesson Title: Steer-a-Shape
Concept: Geometrical Shapes
Grade Level: PK-1st
Source: Charles, Chancellor, Moore, Schielack, & Van de Walle. (1999). *MATH*. Scott Foresman-Addison Wesley Longman, Inc., pp. 161A & 163A. Idea adapted by Heidi Huffer and Niki Skinner (inservice teachers)
Materials: Colored tape, worktable, small toy cars and trucks, gift bag, tissue paper/tracing paper, markers.

5E PHASES

Prior to the lesson, use colored tape to make a square, circle, triangle, or rectangle on each child's worktable. The tape will serve as the outside edge of each shape. Place small toys and trucks in a decorated surprise bag or box.

Engagement (Motivation):
- Pass out one sheet of plain paper and a marker to each child.
- Ask each child to put a dot on his/her paper. Tell the children to start at the dot and, keeping the point of the marker on the paper at all times, begin drawing with their markers. When you say, "Stop," the children then must move the tip of their marker along their paper to the dot where they started.
- Ask children, "What can you tell me about what you have on your paper?" Listen for words such as "lines," "points," "circle," "curves," "square," etc.
- After several children share, have children exchange their papers with a classmate. With the caps on their markers, each child should start at the dot and trace the classmate's drawing.

Exploration (Hands-on, Minds-on Concept Exploration):
- Ask each child to use his/her fingers to trace the tape on top of their worktables.
- Next, direct the children's attention to the decorated container and ask, "Who can guess what's in my special container?"
- Have each child reach into the container, without looking, and retrieve a car or truck.
- Give the children time to "drive" their vehicles around on their workspaces for a few minutes.
- Have children place their vehicles on the tape. Tell them, "When I say 'go,' you can drive your vehicle but it must stay on the tape. Make sure you stop your vehicle where you started."
- Have children switch workspaces so that every child has an opportunity to experience driving on each shape. When the activity is completed, have children place their vehicles to the side.

Explanation (Concept Development):
- Initiate a discussion by allowing children to share what they experienced while driving their vehicles on the tape. Children will begin to perceive differences and similarities in shapes as they compare their experiences with their classmates.
- Ask children to describe what was taped onto their workspaces, listening for terms

such as "curve," "lines," "points," and "circles," and for comments about shapes and number of sides, etc.

Expansion (Concept Relevancy):
- Provide each child with a piece of tissue/tracing paper, and instruct the children to carefully trace their shapes with a marker.
- When finished, ask the children to carefully drive their vehicle on their traced shape.
- Ask the children to describe what would happen if they drove a real car/truck on a real street that was in the shape of a circle, square, triangle, etc.
- Seek children's stories about steering their tricycles or bicycles around their neighborhood. Ask questions such as, "What shapes did you make with your tricycles/bicycles?," "Can you think of any shapes that your neighborhood roads have?," "What shape does the road have where you live?," and "Which shape would you most like to go for a ride on, and why?"

Evaluation (Formative/Summative Assessment):
- Ask various questions and initiate discussions throughout the lesson, during both individual and large-group activities.
- When appropriate, record comments (anecdotal records) in a log to indicate the level of success each child experienced with shapes.
- If they have writing skills, some children could answer your questions in their journals. They also might choose to draw pictures.
- Children might draw pictures of their neighborhoods or classroom. They can share their pictures with their classmates and describe what shapes they see in their drawings. Children can place their "shape" drawings in their portfolios.

SAMPLE LESSON

Lesson Title: Can You See With Your Hands?
Concept: Observation, classification, comparison; senses
Grade Level: K-2nd
Source: Cooney et al. (2000). *Scott Foresman Science*. Addison-Wesley Educational Publishers, pp. D6-D7.
               Idea adapted by Caroline Hern, Lyndsey Pugliese, and Angie Thielsen (preservice teachers)
Materials: Paper; pencils; assortment of buttons (many shapes, sizes, and colors); blindfolds (sunglasses with construction paper covering lenses); cotton balls.

5E PHASES

Engagement (Motivation):
- Direct children to close their eyes and put their hands out in front of them.
- Place a cotton ball in each child's hands and instruct the children to interact with the object.
- Initiate children's responses by asking, "What can you tell me about the object?" and "Describe some of the characteristics you notice about the object in your hands."
- Based on the children's responses, create a classification chart on the board. Emphasize physical properties/characteristics such as texture, size, and shape in relation to the cotton ball. After a brief discussion of various characteristics, encourage the children to guess what the object is.
- Then initiate a discussion by asking children what their sense of touch can tell them.

Exploration (Hands-on, Minds-on Concept Exploration):
- Assign children in pairs to explore more objects. Designate a materials manager from each pair.
- Explain the materials and safety responsibilities to the children. (Each materials manager is responsible for picking up a blindfold, paper, and pencil for his or her group.)
- Explain that a container of objects (buttons) will be placed in front of each pair, and that they are going to explore the objects just as they did earlier with the cotton ball. The students who are not acting as the materials manager put the blindfolds on. Explain that the blindfolded partners are going to be responsible for saying all of their ideas aloud so that the materials managers can then record them. At this point, ONLY the blindfolded partners are making observations about the objects.
- Provide each pair of children with a container of buttons. Instruct the materials managers to open the container and place it in front of their partners.
- Have the children explore the objects in front of them and describe to their partners how the objects feel. "What characteristics do you recognize as you feel the objects? Think about similarities and differences in the objects." The material managers record their partners' thoughts in writing or by drawing a picture on paper. (During this time, observe the children's interactions and listen to their observations about the objects.)

Explanation (Concept Development):
- Initiate a discussion about the children's findings. As children report their ideas/characteristics, list them on the board.
- Review the children's contributions that are written on the board and emphasize the classifications and observations made.
- Initiate further discussion by asking children how they experienced exploring the buttons without their sense of sight. Ask, "How many of you were surprised when you actually saw what the objects were?" and "What did you think that the objects were when you were using only your sense of touch?"
- Provide an opportunity for children to share their experiences with the buttons and address any additional questions.

Expansion (Concept Relevancy):
- Now inform the materials managers to put on the blindfolds. When ready, the blindfolded children must organize the buttons into groups according to the characteristics that their partners choose, such as size, shape, etc. The non-blindfolded children will check for accuracy and challenge their partners to group the buttons based on a variety of characteristics.
- As time permits, the pair of children can take turns being the classifier and the person telling his/her partner how to classify.
- For further connections, children can be encouraged to find five objects in their homes (with adult help to ensure safety) and place them on a table. Using some or all of the objects, the child should classify them based on a common physical characteristic, such as color, shape, size, texture, etc. The child then could ask a participating adult what the items have in common. Example: "I put the candle, quarter, and picture frame together. Can you guess what they all have in common?" (circle shape). You can send a letter home encouraging adult family members to gather objects they can classify according to their choice of characteristics. The child would guess what the adults' chosen objects have in common.

Evaluation (Formative/Summative Assessment):
- Ask various questions and initiate discussions throughout the lesson, during both partner and whole-class activities. When appropriate, record comments (anecdotal records) in a log to indicate the level of success each child experienced with classification. Through observation, you can use a checklist or rating scale to document what concepts the children are grasping during the lesson.
- Following the home activity, encourage children to draw a picture of, or write in a journal about, the objects they used. During a class discussion, children can play a guessing game in which they stand before the class, one at a time, and show their pictures or read their journal samples while their classmates guess what their objects have in common.

SAMPLE LESSON

Lesson Title: Exploring Teeth
Concept: Characteristics of teeth in relation to food eaten
Grade Level: 2nd-3rd
Source: Daniel et al. (2000). *Science*. McGraw-Hill School Division, pp. T69C & D.
                Idea adapted by Jami Davis, Amanda Schaefer, and Amanda Stiles (preservice teachers)
Materials: Resealable plastic bags of beef jerky and lettuce for each child, leaves, crackers, forks, rocks (rounded and rough), and newspaper.

5E PHASES
Prior to the lesson, prepare two resealable plastic bags for each child, one with a piece of beef jerky and one with lettuce. (Check children's records for allergies or special conditions before choosing the food items for this activity.) Since food will be included in this lesson, children must wash their hands prior to the lesson.

Engagement (Motivation):
- Begin the lesson by holding up two plastic storage bags, one containing a piece of beef jerky and one containing lettuce.
- Inform children that they will each receive one bag of both food items. (Proper handling of food and appropriate behavior will be discussed at this time.)
- Ask the children what senses they could use when investigating the two food items. Encourage children to address how all five senses can be used with food.
- Pass out two bags to each child and have them enjoy the items, using all five of their senses.
- After children explore the food items by using their senses, the class shares what they discovered through discussion.

Exploration (Hands-on, Minds-on Concept Exploration):
- Provide each child with enough newspaper to cover their desks and inform them that they each will receive two leaves, two crackers, two forks, and two rocks (rounded and rough).
- Discuss appropriate behavior and safety rules before distributing the materials.
- Let the children know that they will be using only four of their senses—touch, smell, sight, and hearing—during this investigation.
- Pass out the supplies to the children.
- Explain that the rocks and forks are tools that can be used during their exploration for tearing and grinding. Allow a volunteer to demonstrate grinding and tearing motions.
- Allow time for children to explore the materials fully.

Explanation (Concept Development):
- Ask children to share what they discovered during their explorations.
- Record children's ideas and contributions on the board. If necessary, guide the discussion by asking, "What happened when you used the rock/fork with the cracker/leaf?"
- Lead children in sharing how the tools that they used worked in a similar way to teeth.

Ask questions such as, "What teeth are sharp like the fork?" and "What teeth are flat like the rock?" Children eventually should determine that the fork is sharp like canine teeth, which are needed for tearing, and that the rock is flat like molars and incisors, which are for grinding.
- Pass out mirrors so that children can observe the characteristics of their own teeth. Allow them to share what they have discovered. Encourage children to connect the tools' (fork and rock) characteristics to their own teeth.

Expansion (Concept Relevancy):
- Walk around the room holding a shark's tooth (or a tooth from a similar animal) and give each child a chance to observe its characteristics. Do not tell the children anything about the tooth. Ask children to keep their thoughts to themselves until everyone has had an opportunity to observe the tooth. While they are waiting, ask them to think about what this particular tooth was used for and where it came from.
- After all children have observed the tooth, ask for volunteers to describe what they saw in your hand.
- Continue the discussion by asking children what they think the tooth was used for. Encourage children to use the terms "tear" and "grind." Also encourage children to describe what types of food are eaten by animals with teeth that tear and teeth that grind.

OPTION: If an animal's tooth is not available, begin by explaining that you have a picture of some teeth and that you need help in discovering what the teeth are used for. Share pictures of teeth from several different kinds of animals. Cover up the image of the animal except for its teeth, and allow children to discuss what the animal might be and what kind of food it would eat. Continue with several examples of sharp teeth, flat teeth, and a set of teeth that includes both types.

Evaluation (Formative/Summative Assessment):
- Ask various questions and initiate discussions throughout the lesson, during both individual and large-group activities.
- Design a checklist or rating scale that includes criteria for both science process skills and science attitudes.
- After the discussion is completed in the Expansion phase, children could create one or two animals with teeth that cut and teeth that grind. Children should draw their animal in an appropriate habitat and/or write a story about their animal. They should show or describe how and what their animal eats. When completed, children should be encouraged to share their creations and stories with their classmates. Children's drawings and stories could be displayed on a bulletin board titled "Exploring Teeth," and later included in their portfolios.

SAMPLE LESSON

Lesson Title: Awesome Air
Concept: Moving air has lower pressure
Grade Level: 3rd and up
Source: Emily A. Johnson, Ed.D., Assistant Professor of Early Childhood Education, Kennesaw State University, GA.
Materials: Two sheets of notebook paper per child, flexible drinking straws (one per child), cheese balls (one per child), Ping-Pong balls (one per child), cone-shaped paper drinking cups with the tip removed (one per child), vacuum cleaner hose, bits of torn paper, individual journals.

5E PHASES

Engagement (Motivation):
- Direct children to take two sheets of notebook paper and hold them about six inches apart, parallel to one another and perpendicular to their faces.
- Ask the children to predict what will happen to the paper when they blow between the sheets.
- Give them the opportunity to blow between the sheets and observe the results. (Most children predict that the paper will blow apart, but instead the paper pulls inward.)
- Initiate children's responses by asking, "What happened to the papers?" and "Why do you think that it happened?"

Exploration (Hands-on, Minds-on Concept Exploration): (Prior to distributing materials, the teacher will review appropriate behavior and required safety measures)
- Give children flexible drinking straws and instruct them to bend them into an L shape. Have them place the straws in their mouths, with the long end of the L pointed toward the ceiling, then blow through the straw and feel the air as it comes out the end. They are to make observations about the air, such as temperature and speed of airflow.
- Distribute a cheese ball to each child; instruct them to blow through the straw with a forceful and consistent strength while placing the cheese ball into the stream of air. (The cheese ball will be held in place as long as air is being blown forcefully through the straw.)
- Children investigate the phenomenon and write about their discoveries in their journals. Additional cheese balls may need to be provided as some get lost or crushed.
- Collect all used materials in the garbage.

Explanation (Concept Development):
- Initiate a discussion about the children's findings. Children report their ideas about why the cheese ball stayed aloft in the stream of air.
- Explain that two forces are at work in these two different experiences: low pressure and high pressure. Lead the children to determine that low pressure was created between the two pieces of paper when they blew through them. Then ask, "If low pressure is where there is moving air, where was the low pressure with the straw and cheese ball?"
- Explain that moving air creates low pressure and still air creates higher pressure. Guide

the children toward determining that the still air had higher pressure and it held the cheese ball in place. The still air had higher pressure and it pushed in the sides of the paper.

Expansion (Concept Relevancy): (Prior to distributing materials, review appropriate behavior and required safety measures.)
- Explain that each child will receive a Ping-Pong ball and a cone-shaped paper drinking cup with the tip cut off. The children will put the tip of their cups into their mouths while holding their heads up and looking at the ceiling. They then blow through the cone and feel the air inside the cone. Ask the children to predict what will happen to the ping-pong ball when it is placed in the cone and they then blow forcefully into the cone. (The natural inclination is to say that they can blow the Ping-Pong ball out of the cup; if the children apply what they observed from the straw and cheese ball and the paper experiences, however, they will accurately predict that the Ping-Pong ball will stay in the cup.)
- Children investigate with the Ping-Pong balls and paper cones and record their discoveries in their journals. They can experiment by holding their heads back and blowing into the cup while the Ping-Pong ball is in the cup. They also can hold their heads down and gently hold the Ping-Pong ball in place while blowing forcefully.
- Ask the children to share real-life situations related to air movement that connect to the discovered concept.
- Explain that the moving air in a tornado causes low pressure, which creates lift. Demonstrate this concept by placing torn bits of paper on the floor. A volunteer holds a vacuum cleaner hose very close to the torn bits of paper while the teacher vigorously swings the top of the hose in a circle. Low pressure is created at the top of the hose and the bits of paper are forced up the hose by the high pressure at the bottom. Bits of paper are slung out of the top of the hose and all over the room.
- Ask the children to share what careers would involve knowledge of air movement. STS applications also could be discussed.

Evaluation (Formative/Summative Assessment):
- Ask various questions and initiate discussions throughout the lesson. A checklist can be used to determine what concepts the children are grasping during the lesson.
- Ask children to draw a picture of the straw and cheese ball experiment, and label where the air is moving the fastest, where there is low pressure, and where there is high pressure. Children also can write in their journals about their discoveries and any unexpected outcomes. The same can be done for the cone and Ping-Pong ball experiment.

Benefits of the 5E Learning Cycle Format

Numerous benefits result when early childhood learners participate in the learning cycle:

Cognitive Benefits:
- Caters to the learning needs of children through exploration and discovery, while promoting disequilibrium and fostering assimilation and accommodation of knowledge
- Allows children to construct their own knowledge in line with developmentally appropriate practice theories as they build on their prior knowledge
- Allows children to reinvent knowledge on their own while taking ownership of knowledge, by utilizing both process skills and math and science content
- Provides concept connections for children that are found in various subject areas
- Emphasizes real-life contexts for concept relevancy and STS, while promoting career and job awareness
- Uses alternative assessment methods, versus only paper-and-pencil tests.

Affective Benefits:
- Arouses, motivates, and capitalizes on children's natural sense of curiosity based on prior knowledge
- Involves children actively, rather than passively, in hands-on, minds-on learning
- Provides learner-centered opportunities for children to extend learning beyond the basic lesson goals/objectives, demonstrating a respect for children's knowledge
- Promotes learning and strengthens learners' confidence about content and process skills
- Allows children to construct concepts cooperatively, emphasizing socialization skills and the value of various perspectives while assimilating knowledge.

Summary

The learning cycle lesson format originally used in the SCIS program resulted in high achievement, both cognitively and affectively, for children. Teachers who implemented the 5E Learning Cycle model in early childhood classroom settings made the following comments:

- Children's potential was revealed as children took the lesson further than anticipated.
- Children are able to think about their learning, instead of being told what they will be learning.
- Children are engaged in the process of inquiry.
- The approach expands children's thinking and lets them come up with answers, rather than being "taught."
- I love how engaged the children get and how well the classroom management goes when children do this model.
- The model builds on children's natural curiosity, which increases their motivation to learn.
- Children are motivated to jump-start their thinking, conduct an investigation, examine their findings, and expand and extend what they just learned. The model engages the learner and allows them to use many learning styles.
- Children manipulate materials and are in charge of their own learning, with guidance from their teacher.
- The model creates a balance in discovery learning and inquiry.

Clearly, this engaging teaching/learning model should find its way into every early childhood classroom setting.

References

Charles, R. I., Chancellor, D., Moore, D., Schielack, J. F., & Van de Walle, J. (1999). *MATH - Grade K. Vol. 2*. Glenview, IL: Scott Foresman-Addison Wesley Longman.

Charlesworth, R., & Lind, K. K. (2003). *Math and science for children* (4th ed.). Albany, NY: Delmar.

Cooney, T., DiSpezio, M. A., Foots, B. K., Matamoros, A. L., Nyquist, K. B., & Ostlund, K. L. (2000). *Scott Foresman Science – Grade 1*. Glenview, IL: Scott Foresman-Addison Wesley Educational Publishers.

Daniel, L. H., Hacket, J., Moyer, R. H., Baptiste, H. P., Stryker, P., & Vasquez, J. A. (2000). *Science – Grade 2*. New York: McGraw-Hill School Division.

Gallenstein, N. L. (1999). Making mathematics meaningful. *Reaching Through Teaching, XII*(2), 13-14.

Jackson, A. L., Johnson, M. L., Leinwand, S. L., Lodholz, R. D., Musser, G. L., & Secada, W. G. (1996). *Mathematics in action—Grade K*. New York: Macmillan/McGraw-Hill School Division.

Kellough, R. D., Carin, A. A., Seefeldt, C., Barbour, N., & Souviney, R. J. (1996). *Integrating mathematics and science for kindergarten and primary children*. Englewood Cliffs, NJ: Prentice-Hall, Inc.

Martin, R., Sexton, C., & Gerlovich, J. (2001). *Teaching science for all children* (3rd ed.) Needham Heights, MA: Allyn and Bacon.

Martorella, P. H. (1998). *Social studies for elementary children: Developing young citizens* (2nd ed.). Upper Saddle River, NJ: Prentice-Hall.

National Council of Teachers of Mathematics. (2000). *Principles and standards for school mathematics*. Reston, VA: Author.

National Research Council. (1996). *National science education standards*. Washington, DC: National Academy Press.

Sheffield, L. J., & Cruikshank, D. E. (1996). *Teaching and learning elementary and middle school mathematics* (3rd ed.). Englewood Cliffs, NJ: Prentice-Hall.

Winthrop, E. (1986). *Shoes*. New York: HarperCollins.

Chapter Three
Concept Attainment Methods for Early Childhood Learners

"Let's do another one!" "What else is in the mystery box?" "I want to make one too!" These are typical comments from children who have experienced the excitement of participating in concept attainment models. By exploring concepts through active engagement, learners of all ages gain experience in thinking inductively as they classify information.

Concept Formation

We use concepts to organize our world; they can act as a mental "filing cabinet." We file similar concepts/ideas together, forming categories by observing and identifying similarities and differences of objects/ideas. We start by learning broad categories, then proceed from the general to the specific.

The process of elimination allows us to determine that some things are included in a category (examples), while others are not (non-examples). Even very young children are capable of performing this process. It is always very exciting for parents when their child begins to acquire a vocabulary. As they learn words, children are using the process of elimination. For example, a baby's first spoken word might be "Da Da." Initially, the baby might point to and identify each person he/she encounters as Da Da. Eventually, Da Da might be narrowed down to apply only to males, eliminating females. With additional exposure to a variety of males, and feedback from the adults, the baby eventually learns that the word Da Da belongs to one very special male. Another example is a child's use of the word "dog." We are delighted when the child correctly identifies the family pet as "dog." Yet, while driving down the road and through the country, the young child may identify as a dog *all* four-legged animals observed, whether it be a horse, cow, pig, cat, llama, goat, etc. Again, with a variety of experiences, repetition, and practice, the young child will use the process of elimination and eventually understand that a dog is a specific type of furry four-legged animal that is often a pet and has a set of particular characteristics—usually a wet nose, wagging tail, and a bark.

Exposure to examples and non-examples of a concept help children learn what something is and what something is not. Conflict and contradiction are necessary to the theory-building process when discovering concepts (Chaillé & Britain, 1997). As children grasp the concepts, they are assimilating and accommodating structures of knowledge. "A successfully organized group of objects provides them with immediate satisfaction that the task is over and is done well" (Winnett et al., 1996, p. 11).

As previously mentioned in Chapter One, the basic math concepts of comparing, classifying, and measuring are also basic process skills in science. The process skills most appropriate for preschool and primary children are observation, comparison, measurement, classification, and communication. These skills will be useful later as children learn to gather, organize, and record data; infer relationships; predict outcomes; hypothesize; and identify and control variables (Charlesworth & Lind, 2003). The child's developmental level and the learning experiences provided by the teacher will determine the range of process skills that a child will successfully experience. When children discover and attain concepts through appropriately designed learning experiences, they are processing information through observation, comparison, classification, and communication, and by creating hypotheses. The methods for discovering and attaining concepts presented in this chapter will provide children with opportunities to be actively engaged in the creative construction of knowledge by using basic mathematics and science concepts and process skills.

Classification

Classification is a basic concept experienced at the early childhood level in both mathematics and science. Classification activities help children focus their observation skills and develop a strong understanding of the concepts "alike" and "different." Classification also provides children with opportunities to reason, problem solve, make decisions, and be in control of their learning.

When participating in the attainment of concepts through discovery, children must have an understanding of classification. Early childhood math and science offers many opportunities for children to group and sort objects according to categories. First, children must use their senses as they observe a set of objects or ideas. While observing, children look for similarities and differences as they compare those objects/ideas. Classification requires children to put together objects/ideas that have one or more characteristic in common. Children learn that objects can be grouped together according to various common features/attributes.

Children at the early childhood level need many opportunities to explore, observe, describe, select, estimate, and group a variety of objects. The ability to classify is a fundamental skill for children in understanding whole number (Kellough et al., 1996), and it helps children to organize their world. A variety of classification experiences can be presented to children for the exploration of length, area, volume, weight, color, shape, geometric features, and texture. Charlesworth and Lind (2003, pp. 134-135) provide the following groupings for children to use during math and science classification activities (in play and formal instruction):

Color:	Items that contain or are all the same color.
Shape:	Items that are round, square, triangular, etc.
Size:	Items that are big, small, fat, thin, short, tall, wide, narrow. (It is important to remember that a reference item should be used in order for children to determine how items should be grouped. For example, what is considered to be big to one child may be considered small by another.)
Material:	Items made of wood, plastic, glass, paper, cloth, and metal.
Pattern:	Items that have different visual patterns, such as stripes, dots, flowers, or plain (no design).
Texture:	Items that are smooth, rough, soft, hard, wet, dry, sticky, etc.
Function:	Items that do the same thing or are used for the same thing (all are for eating, writing, playing music, etc.).

Association: Items that do a job together (candle and match, milk and glass, shoe and foot), come from the same place (bought at the store or seen in the zoo), or belong to a special person (the hose, truck, and hat belong to the firefighter).
Class name: Words/names that may belong to several items ("people," "animals," "food," "vehicles," "clothing," "homes," etc.).
Common features: Items that have some element (e.g., handles, windows, doors, legs, wheels, etc.) in common.
Number: Items that are grouped into specific amounts, such as pairs; groups of three, four, five; and so on. (Older children can classify sets of numbers as odd or even, prime or composite, divisible by 5 or not, etc.)

When presenting children with opportunities to discover and attain concepts, early childhood educators must remember to begin with concrete objects, then move to the pictorial level (objects with pictures or pictures with words), and then progress to the symbolic level (words or numerals). This text offers concept attainment examples for all three levels of learning.

As previously mentioned, Jerome Bruner has explored methods that stimulate and encourage children's thinking in the classroom through inquiry and discovery. Bruner believes that both the teacher and child must become involved in the process of learning (Howe & Jones, 1993). As children are very interested in things that pique their curiosity and appear mysterious or puzzling, concept attainment activities include an element of mystery. Children take on the role of detective as they solve concept attainment activities in stimulating learning environments. In doing so, they take a newly introduced object or idea and place it in some category that they have previously discovered or identified (Kellough et al., 1996). Knowledge is built upon knowledge as children recall past experiences, reflect on them, and add new experiences to them. When participating in the attainment of concepts, children must observe carefully and be aware of their own thought processes.

Early childhood educators should encourage self-directed problem solving and experimentation with objects and ideas. A learner attempting to solve a challenging situation follows several "thinking" steps; similar steps lead a child toward discovering and attaining concepts. First, the learner must recognize that a gap (problem) exists between his/her current understanding/knowledge and desired understanding/knowledge. Second, the learner gathers information in order to develop tentative solutions (hypotheses) to the situation. Third, the learner tests the solutions using current knowledge. Fourth, the learner gathers additional data in order to confirm, check, or verify hypotheses. Fifth, after testing solutions, the learner draws conclusions and plans to apply the new knowledge (Kellough et al., 1996).

When presenting group concept attainment activities, educators, through questioning, should monitor these steps carefully. Once children are familiar with the thinking processes involved, concept attainment activities can be placed in learning centers for children to tackle individually or as members of a small group.

The discovery and attainment of concepts can be used effectively across all subject areas and grade levels. This chapter contains examples of how early childhood educators can promote methods for discovering and attaining concepts when teaching mathematics and science to children, and a discussion of the benefits of allowing children to discover and attain concepts on their own. Readers also will find various adaptations appropriate for the early childhood level, including suggestions for implementation. The following suggested methods for attaining mathematics and science concepts are in line with how learners define knowledge.

Methods for Attaining Concepts

Children of all ages enjoy immensely the first method presented for attaining concepts, which is based on a very effective model that is presented in the book *Models of Teaching* by Joyce, Weil, and Calhoun (2000). This adaptation, while very similar to the model presented in *Models of Teaching*, is focused specifically on reaching early childhood-age children (ages 3-8). Suggestions are included for presenting this model to both preservice/inservice teachers and children.

When presenting this model to preservice/inservice teachers, I recommend following Bruner's stages of knowledge representation. Remember the importance of working with children first on the concrete level (actions on objects), then moving to the pictorial level (pictures of objects), and finally to the symbolic level (words, numerals, signs).

When introducing a concept, teachers first assess children's current understandings and then create additional lessons/activities appropriate for the concept that are based on children's levels of understanding. Such concept attainment models also can be used as effective assessment tools during lessons or to conclude lessons/units on a particular concept. Additionally, concept attainment models can be very effective when reinforcing concepts previously taught and connecting them to additional concepts.

In the first set of concept attainment models provided here, the following steps are recommended when first conducting the model; they can be adjusted according to your children's level and your own comfort.

Teacher's Role (Planning the activity):

- Determine the concept that children will attain.
- Clarify the concept by defining it.
- Choose objects/pictures/symbols/words, etc. that children can identify as examples (yeses/"thumbs up") or non-examples (nos/"thumbs down"). Approximately 10-15 "yes" and 10-15 "no" items should be determined in advance.
- Arrange examples and non-examples in order for presentation. Items should be arranged so that, initially, more than one concept might apply. This will be demonstrated during the concept attainment activities. As more "yes" items are presented, children eventually will narrow their ideas to one concept. At the early childhood level, I recommend that the teacher always start the activity by presenting two very obvious "yeses," and then two "nos." "Yes" items all must have a common attribute/feature. "No" items should contain a variety of attributes, as opposed to all "no" items simply being the opposite of the "yes" items. (Refer to Figure 3 – Farm Animals. The characters are from the Inspiration® 6.0 computer program.) Children at this level need to observe "yes" items carefully in order to compare and contrast with the "no" items for appropriate classification.
- Review items in advance for presentation and anticipate children's reactions. Be certain that every "no" item to be presented *will not* fit the concept/idea. If necessary, adjust the order of items. Although a written plan should be developed, it is important to remember that adjustments can be made to the order of items presented while implementing the activity. *Be flexible—every group of children will respond differently to the same activity!*
- Determine when the concept attainment model should be presented in the curriculum for learning effectiveness. For example, would it be most effective as an opening activity, an activity within the lesson, or a closing activity?

Children's Role (Discovering the concept):

- Children carefully observe the items that the teacher presents, and try to determine what the teacher's idea is based on. They also need to discern what attribute/feature the "yes" items all have in common that the "no" items do not have.
- Children sit quietly and keep their ideas to themselves so that their classmates can think on their own about what the idea might be. "Be a quiet detective and keep the secret to yourself." Let children know that it is okay to put their hands over their mouths if their answer otherwise might slip out. At times, it might be necessary to remind children that "This is a silent model!"
- When the concept is finally attained, children can share the other ideas/concepts that came to mind throughout the model as "yeses" and "nos" were presented. Ask children to describe why particular ideas came to mind, why they discarded certain ideas, and what made them decide on their final solution. Also, have children share the feelings they experienced when they were sure that they knew the teacher's idea and how they felt when they were not sure.

Model Implementation

To begin, be certain that all children are able to see the presented items throughout the activity, whether they are seated at their desks or on the floor in front of you. Describe the concept attainment activity for the children. They will be responsible for determining what your idea is, based on the "yes" and "no" items that you will present to them. The "yes" items all have something alike/similar/in common—your "idea." The "no" items will not fit with the "yes" items; they do not fit with your idea. Tell the children that at first they might think of more than one idea that could connect the "yes" items together, but that as they observe more "yes" and "no" items they probably will determine that only one idea works.

Farm Animals

Yes | No

Figure 3

Inform the children that they should act like secret detectives and keep their ideas to themselves, for everyone deserves an opportunity to think about and determine the secret/idea on their own. Ask for a volunteer to paraphrase the directions for the activity. Be aware that when first presenting this model, you will probably need to remind children often of their role. As they continue to practice the model, they will know and value their role.

To begin the activity, hold up one "yes" item from your mystery bag/box and ask the children to look at it carefully. After they have carefully observed the item, explain that you are going to put all of the "yes" items (examples) by the word that spells "YES" (a thumbs up sign also would be appropriate). Be certain to place all of the items where children will be able to see them throughout the activity. Hold up another "yes" item (example). Ask the children to observe both "yes" items carefully and think about how they are alike/same. "Think about what they both have in common. Keep your thoughts to yourself." Then, ask for a show of hands as to how many children have at least one idea on how both "yeses" are alike. "Who has two ideas? Three ideas? More than three ideas?" This step will indicate how many children are actively thinking. Plus, it acknowledges the importance of their involvement.

Next, hold up a "no" item. Inform the children that this item does NOT fit with the "yes" items: "It does not fit my idea." Give the children time to carefully consider what feature(s) the "yes" items have in common that the "no" item does not have. Remind them that this is a SILENT model. As detectives, they must keep the secret to themselves. If they share their thoughts, they may deny other children the chance to experience the thinking process. Through these reminders, you also will be discouraging learned helplessness for those children who might prefer to have someone else do the thinking. Then, hold up another "no" item and remind children that this item will not fit with the "yeses."

Point to the two "yes" items and remind children to compare them to the two "no" items. "Think about what the 'yeses' have in common that the 'nos' do not have. Who thinks they have my idea?" Ask the children for a show of hands. "Who has more than one idea? Two ideas? Three ideas?" If it appears that many children think they know your idea, then proceed to the next step; if not, hold up another "yes" example and allow more time for the children to observe and think about what all of the "yes" items have in common. Depending on where you sense the children to be, continue to hold up "yeses" and "nos" until most of them have at least one idea on how the "yeses" are alike.

Next, hold up an item from your mystery bag/box and ask for a show of hands as to how many children think that this item belongs in the "yes" group or the "no" group, or if they are "undecided" about where this item fits. Using the word "undecided" lets children know that it is okay if they have not yet firmed up their thoughts; they are still detectives. The teacher should watch carefully for when children raise their hands. Children's responses are indicators of how they are progressing in attaining the concept. Plus, their responses provide a signal as to what the teacher should do next.

Inform the children that the item you are now holding is a "yes." Place the item with the other "yes" items and again remind the children to carefully observe what all of the "yeses" have in common that the "nos" do not have: "Think how the 'yeses' are alike." Ask them if anyone had to change their idea when you placed that item with the "yeses." Sometimes when children learn that what they thought would be a "yes" or "no" item is instead the opposite, you will hear a gasp of frustration. The children are experiencing what Piaget termed "disequilibrium"—they are thrown off-balance because they cannot assimilate the item into their present scheme. They will want to achieve balance (equilibration), and thus will strive toward accommodating the new knowledge. As previously mentioned, Piaget encouraged teachers to create disequilibrium in order to stimulate learning.

Continue to pull out items from your mystery bag/box and ask children by a show of hands whether they think each item is a "yes" or a "no," or if they are "undecided." After they respond, inform them as to whether the item is a "yes" or a "no," and place it in the correct group. Then ask the children, "How many think that you have my idea?" Inform them that you are now going to walk around the room and let them whisper their secret idea to you. Ask the children to raise their hands if they want to whisper their idea to you. If the children are writers, they may choose to have you look at what they wrote on a piece of paper. Tell the children that you will not yet let them know if their idea is the same as your idea. You just want to know what they are thinking. I have found that when the teacher lets children know whether or not they have identified the idea, those who are correct often begin to lose interest in the activity. Those who are incorrect often become frustrated. Instead, just listen at this stage. Knowing what the children's ideas are will help you choose what items to present next.

If most children do not have your idea, show more "yes" and "no" items, one at a time, again asking them by a show of hands if they think each item is a "yes" or a "no," or if they are "undecided." When holding up items, be very careful not to say, "This item will really help you figure out my idea." If the item does not help the children clarify the concept, they often become even more frustrated. It is important to remain neutral when presenting items.

At this point, if you believe most children have your idea, ask someone to look around the classroom and name an item that they think is also a "yes" item. Ask the rest of the class by a show of hands whether they think that the child's suggested item is a "yes" item or a "no" item, or if they are "undecided." Repeat this activity several times with different volunteers. If possible, after the children respond to the item, place the suggested item in the correct category in front of them so they can view it. If a suggested item is permanently fixed in its location, it would be appropriate to place a "yes" or "no" label on it. Some children will ask if they can choose a "no" item. I suggest that you initially focus on "yes" items in order to help others determine the attribute/feature that the "yeses" all have in common, rather than what all the "nos" do not have. Just use your discretion. At this time, ask the children again how many think they have your idea. Walk around the room again and allow them to whisper their idea to you. Remember—do *not* yet inform them whether their idea is the same as yours.

When all of the children, or the majority, have your idea, then it is time for the detectives to share. Say, "On the count of three, I want all of you to say what you think my idea is: one, two, three." By allowing children to respond chorally rather than asking for only one child to volunteer what he/she thinks the idea is, all children's thinking is acknowledged and valued. Plus, the children who are still not quite certain of the concept will feel less threatened. At this time, let the children know what your idea was. It should be the same as the children's choral response. *Always* ask the children if anyone still has another idea that fits the "yeses" and "nos" that you presented. This can definitely happen. If so, allow each child to share why his/her idea would also apply. Sometimes, when children think that their concepts also apply to the presented examples and non-examples, there will be one or more items in the "no" category that will not fit their concepts. Gently help them discover for themselves how their idea does not apply to the presented examples/non-examples. Praise the children for their "good thinking."

Debriefing
Ask the children to share all the ideas they had during the course of the activity. It is always very interesting to hear the variety of their ideas; the more diverse the class, the greater the variety of ideas will be. Children and teacher alike benefit from hearing alternative perspectives. Ask the children to share how and why their ideas changed as they progressed throughout the activity. Again, their comments typically are very interesting. Finally, ask

the children how they felt when they thought they had your idea. They usually say that it made them feel good or smart. Ask the children to share how it felt when they were "undecided." Some children say that they felt sad or not smart because some children knew, but they did not. Preschool and kindergarten children often appear to become less frustrated with learning than do some primary-grade children. It seems that an emphasis on doing something "right" plays a big factor in primary children's attitudes toward their learning. It is important to remind children that prior experiences with the examples and non-examples will differ from individual to individual. That is often the reason why some children hone in on a particular concept sooner than others do. Usually, as children gain experience with the model, they become more patient with themselves and look forward to solving each mystery.

The next section of this chapter includes examples of the above-described model. Although a list of examples and non-examples is provided for each activity, items can vary depending on availability. Following the examples, adaptations to the model are discussed. The examples provided could be used with various ages and grade levels. One very important factor to remember when incorporating concept attainment activities into your curriculum is to meet the developmental levels of your children in reference to the concrete, pictorial, or symbolic levels. All examples included are labeled at their presentation level. Adjustments can be made to the materials in recognition of the various learner levels. At the end of the chapter, readers will find a list of ideas for creating concept attainment activities.

SAMPLE ACTIVITY

Concept: Things in Nature
Definition: Things in nature are materials or objects that come naturally from the earth without human interference.
Grade Level: 1st-3rd
Source: Sarah Bogolo, Katie Mach, Heather Martins, and Brooke Paul (preservice teachers)

Items for Presentation (Concrete example):

<u>Yeses</u>	<u>Nos</u>
seashells	shoe
gemstone rocks	cooking pot
grass	fire starter log
branch/stick	canned beans
berries	necklace
apple	matches
flower	lantern
rocks	artificial feather
bird feather	bottle of water
leaves	beach towel

Presentation Format:
- "Today, you are going to be detectives. Your mission is to try to figure out what concept/idea I am thinking of. I will show you examples of things that fit my idea. They will go in the 'yes' category. Other things that do not fit my idea will go in the 'no' category. You need to figure out what all of the 'yes' things have in common. As detectives, remember that any clues you come up with need to be kept top secret, so think and work quietly by yourself."
- "First, I am going to start out by showing you two 'yes' items." Hold up the **seashells**. "These are a 'yes.' I'm going to place them over here on the left side of the table, where I will place all of the other 'yes' items."
- Hold up the **gemstone rocks**. "These are also a 'yes.' They also will be placed on this side of the table. Look at both of the 'yes' items and think about what they have in common. How are they alike?"
- Hold up the **shoe**. "This is a 'no.' It will be placed on the right side of the table, where all of the other 'nos' will be placed."
- Hold up the **cooking pot**. "This is also a 'no.' What side of the table should it be placed on, the right or the left?" Place the item on the right side of the table and ask children to look carefully at both sets of items and think about what the "yeses" have in common that the "nos" do not have.
- Ask children to raise their hands in response to the questions: "Does anyone have any ideas on what my concept might be? Who has one idea? Two? Three?"
- "Next, I am going to ask you if you think the item I hold up is a 'yes,' a 'no,' or if you are 'undecided.' Make sure to sit quietly while you are thinking. When I ask you questions, just raise your hands."
- Hold up the **grass**. "Who thinks this is a 'yes'?" Pause and request a show of hands. "A 'no'? Who's 'undecided'? This is a 'yes.' I'm going to place it with the other 'yeses.'"

- Hold up the **fire starter log**. "Who thinks this is a 'yes'? A 'no'? Who's 'undecided'? This is a 'no.'"
- Ask, "How many have at least one idea of what my concept is? Two? Three?" Remind children to raise their hands.
- Hold up a **branch/stick**. "Who thinks this is a 'yes'? A 'no'? Who's 'undecided'? This is a 'yes.' I'm going to place it here with the 'yeses.'"
- Hold up the **berries**. "Who thinks this is a 'yes'? A 'no'? Who's 'undecided'? This is a 'yes.'" Everyone should answer. "Should it be placed on the right or left side of the table? Correct! I should place it on the left side with the other 'yeses.'"
- Hold up the **canned beans**. "Who thinks this is a 'yes'? A 'no'? Who's 'undecided'? This is a 'no.' I'm going to place it on the right side of the table with the other 'nos.'"
- Hold up the **apple**. "Who thinks this is a 'yes'? A 'no'? Who's 'undecided'? This is a 'yes.' I am going to place it on the left side of the table with the other 'yeses.'"
- "Now, think about what the 'yeses' have in common that the 'nos' do not have. Think about how your ideas have changed. Who has one idea? Two ideas? Three ideas?"
- Hold up the **flower**. "Who thinks this is a 'yes'? A 'no'? Who's 'undecided'? This is a 'yes.' I'm going to place it on the left side with the other 'yeses.'"
- Hold up the **necklace**. "Who thinks this is a 'yes'? A 'no'? Who's 'undecided'? This is a 'no.' Should I place it on the left or right side of the table? The right side, because this is a 'no.'"
- Ask, "How many think that you have my idea?"
- "I am going to walk around the room and listen to your ideas as you whisper them to me. Raise your hand if you want me to listen to your idea." If the children are writers, they may choose to have you look at what they write on a piece of paper. Remember not to tell them if their idea is right or wrong—just listen to their thinking.
- Ask, "Who sees something in this room that they think would fit with the 'yeses'?" Ask children to raise their hands; call on a child to share a "yes" item. Then ask, "Who thinks his/her item is a 'yes'? A 'no'? Who's 'undecided'?" If the item can be moved, place it in the appropriate category.
- Repeat the preceding step by asking other children to choose an item that they think is a "yes" item in the room. Each time ask, "Who thinks this is a 'yes'? A 'no'? Who's 'undecided'?"
- Hold up the **rocks**. "Who thinks this is a 'yes'? A 'no'? Who's 'undecided'? This is a 'yes.' I'm going to place it on the left side with the other 'yeses.' Look again to see what all of the 'yes' items have that the 'nos' do not have."
- Hold up the **matches**. "Who thinks this is a 'yes'? A 'no'? Who's 'undecided'? This is a 'no.' I'm going to place it with the other 'nos.'"
- Hold up the **lantern**. "Who thinks this is a 'yes'? A 'no'? Who's 'undecided'? This is a 'no.' On which side of the table should I place it?"
- Hold up the **bird feather**. "Who thinks this is a 'yes'? A 'no'? Who's 'undecided'? This is a 'yes.' Where should I place the bird feather?"
- Hold up the **artificial feather**. "Who thinks this is a 'yes'? A 'no'? Who's 'undecided'? This is a 'no.' Where should I place it on the table?"
- Hold up the **bottle of water**. "Who thinks this is a 'yes'? A 'no'? Who's 'undecided'? This is a 'no' (plastic container). I'm going to place it with the other 'nos.'"
- "Now, how many of you think that you have my idea? Who has more than one idea?"
- Ask children to raise their hands if they want to whisper their ideas to you. Walk around the room and listen/look at their ideas. If the majority or all of the children have the idea, ask them to say it together on the count of three. If not, proceed with sharing items from your mystery bag, as well as allowing children to volunteer "yes" items that they see in the classroom.

- "Let's look at some more items to be sure of the idea."
- Hold up a **beach towel**. "Who thinks this is a 'yes'? A 'no'? Who's 'undecided'? This is a 'no.' I'm going to place it on the right side of the table with the other 'nos.' "
- Hold up the **leaves**. "Who thinks this is a 'yes'? A 'no'? Who's 'undecided'? This is a 'yes.' Where should I place the leaves? Correct. I will place them on the left side with the other 'yeses.' "
- If necessary, allow children to whisper their ideas to you one more time before you conclude the activity.
- "Now, I will count to three, and I want you to say together what you think my idea is. Okay: one, two, three."

Debriefing:
- Ask if anyone came up with a different idea/concept that would also apply to the items in the activity. Discuss why the concept would or would not work.
- Discuss how everyone came to his or her conclusion about the concept. "What were some of the ideas you had while I was sharing the items?" "What made you change your mind?"
- Take time to discuss the concept and clarify how certain items were indeed "yeses" or "nos."
- Ask children to share the feelings they experienced during the activity.

SAMPLE ACTIVITY

Concept: Objects with numerals on them
Definition: Objects that have numerals printed somewhere on them
Grade Level: 1st-3rd
Source: Lindsay Brewer, Amy Clinton, Laura Hammon, and Amber Lakes
              (preservice teachers)

Items for Presentation (Concrete example):

Yeses	Nos
VHS video in box	sunglasses case
box of colored pencils	film canister (plastic)
box of film	eraser
9-volt battery	journal
book	beaded necklaces
CD	black video (no box)
Ibuprofen pill bottle	mug
box of staples	spoon
measuring cups	stapler
padlock	stuffed bear
protractor	

Order for Presentation:
1. boxed video
2. box of colored pencils
3. sunglasses case
4. film canister
5. box of film
6. battery
7. eraser
8. book
9. journal
10. beaded necklaces
11. black video (no box)
12. CD
13. Ibuprofen pill bottle
14. mug
15. spoon
16. box of staples
17. stapler
18. measuring cups
19. stuffed bear
20. padlock
21. protractor

Presentation Format:
- Begin by introducing the class to two examples that fit the idea. Show the class a **boxed video** and inform them that "this is a 'yes' item" and place it on the left side of the table.
- Introduce the second "yes," which is a **box of colored pencils**. Hold up the box for everyone to see, announce that it is a "yes" item and place it on the left side of the table. Ask the children to think of what the two items have in common. Remind children that they are to keep their ideas to themselves so that everyone has a chance to think on their own.
- Next, introduce two non-examples of the idea. Hold up a **sunglasses case** and announce to the class, "This is a 'no' item" and place it on the right side of the table.
- Introduce the second "no," which will be a black **film canister**. Hold up the container for everyone to see and say, "This is a 'no' item" and place it on the right side of the table.
- Remind children to think of what characteristics the two "yes" items have that are the same, and that the two "no" items will not fit with the "yes" items.
- Ask the children, "How many have one idea? Two ideas? Three ideas? More than three ideas?"
- Next, attempt to create disequilibrium by introducing the class to items with similar

characteristics. The fifth item to be introduced will be a colorful **film box.** Hold up the film box for everyone to see and ask, "Who thinks this is a 'yes'? Who thinks it is a 'no'? Who is 'undecided' about the item?" Provide enough time for children to think and respond by raising their hands. Tell the class that the film box is a "yes," and place it on the left side of the table.
- Introduce a **9-volt battery** to the class, followed by the same three questions: "Who thinks this is a 'yes'? Who thinks it is a 'no'? Who is 'undecided' about the item?" Inform the children that the battery is a "yes," and place it on the left side of the table.
- Ask the children, "Who thinks that they have the idea?" Count the showing of hands.
- Use the three-question technique throughout the presentation for the remaining items. When many hands are raised following the question, "Who thinks they have the idea?," walk around the room and allow children to whisper their ideas to you. (If children are writers, they can write their idea(s) on a piece of paper.) *Do not inform them if an idea is correct or incorrect.* By allowing the children to share, the teacher will have an understanding of where the children are in their thinking. Based on children's ideas, you can then decide what your next action should be.
- Ask for a volunteer to find an item in the room that he or she thinks fits the idea. Again, use the three-question technique when referring to the item chosen by the volunteer: "Who thinks this is a 'yes'? Who thinks it is a 'no'? Who is 'undecided' about the item?"
- Request other volunteers to choose a "yes" item in the classroom; after they do so, ask the three questions.
- Again, by a show of hands, ask how many children think they have the idea. Walk around and listen to, or look at, the children's written ideas.
- When the majority or all of the children have the idea, ask them, on the count of three, to state it chorally.

Debriefing:
- Ask if anyone came up with a different idea/concept that would also apply to the items in the activity. Discuss why the concept would or would not work.
- Discuss how everyone came to his or her conclusion about the concept. "What were some of the ideas you had while the items were shared? What items made you change your mind?"
- Take time to discuss the concept and clarify how certain items were indeed "yeses" or "nos."
- Ask children to share the feelings they experienced during the model.

SAMPLE ACTIVITY

Concept: Objects that have a smell to them
Definition: Objects with a noticeable scent/odor
Grade Level: PK-3rd
Source: Betsy Case, Lisa Klett, and Shelby Surloff (preservice teachers)
Items for Presentation (Concrete level):

Examples	Non-examples
nail polish	notebook paper
perfume	quarter
red cinnamon candle	water bottle
red roses	piece of jewelry
tuna fish	candle holder
ginger spice	stuffed animal
deodorant	pen
apple shower gel	nail clippers
fresh flower	

Presentation Format:
- Opening: "I have a bag full of mystery items. I have a secret idea in my head that you will need to figure out. You will do this by looking at the objects and deciding what they do or do not have in common. It is important to keep your thoughts to yourself, so everyone has an opportunity to figure out the mystery. Please keep your ideas a secret."
- Begin by holding up the **nail polish** and **perfume** and say, "These are both 'yeses.' They have something the same that fits my idea." Tell the children to think of and write down (if they are writers) everything that comes to mind that these two items have in common (same). Place the items together where the children can see them.
- Hold up the **notebook paper** and the **quarter** and say, "These two items are both 'nos.' They will not fit my idea." After the children have looked at the "nos," place them in a different location from the "yeses" for the children to see.
- Hold up both "yeses" and both "nos" and have the children look at all four items. Ask them what they think that the "yeses" have that the "nos" do not have. Ask children, "How many of you have one idea? Two ideas? More than two ideas?"
- Hold up the **water bottle** and say, "This is a 'no.'" Ask the children again how many think they have the idea.
- Hold up the **red cinnamon candle**. "Who thinks this item is a 'yes'? Who thinks it is a 'no'? Who is 'undecided' about the item?" Tell the children that it is a "yes." Always be attentive to how each child responds to each question. Their responses are indicators of where they are in the thinking process, and of how you should proceed with the activity.
- Hold up the **red roses**. "Who thinks this item is a 'yes'? Who thinks it is a 'no'? Who is 'undecided' about the item?" Tell the children that it is a "yes," and place it with the other yeses.
- Review all of the "yes" items again, and then all of the "no" items. "Who thinks they have our idea now? How many of you have one idea? Two ideas? More than two ideas?"

- Hold up the **piece of jewelry** and tell children that it is a "no." "Raise your hand if you think you have my idea."
- Walk around the classroom and allow the children to whisper their ideas to you. Inform the children that you will not tell them whether an idea is your idea, but that you are just going to check with them to see what they are thinking.
- Hold up the **candleholder**. "Who thinks this is a 'yes'? Who thinks it is a 'no'? Who is 'undecided'? This is a 'no.'"
- Ask for a volunteer to point to something in the room that he/she thinks is a "yes" item. Ask the children, "Who thinks the item is a 'yes'? Who thinks it is a 'no'? Who is 'undecided' about the item?" Tell the children if it is a "yes" or a "no" item.
- Show the children the **tuna fish**. "Who thinks this item is a 'yes'? Who thinks it is a 'no'? Who is 'undecided' about the item?" Tell the children that it is a "yes." "Who thinks they have the idea now?"
- Show the children the **ginger spice**. "This is a 'yes' item. Has anyone crossed any ideas off of their list?"
- Show them the **stuffed animal**. "Who thinks this item is a 'yes'? Who thinks it is a 'no'? Who is 'undecided' about the item?" Tell the children that it is a "no."
- If the children need to focus on all of the "yeses" and "nos," take time to review all of the items. Also, if necessary, children may come up to the items and explore them more closely to test their idea (smell).
- Show the children the **deodorant**. "This is a 'yes.' Who thinks they know what the idea is?" Walk around the room and check the children's ideas.
- Ask, "Who thinks that they can show me an item in the room that is a 'no'?" A child volunteers to do so. "Who thinks this item is a 'yes'? Who thinks it is a 'no'? Who is 'undecided' about the item?" Tell the children whether it is a "yes" or a "no" item. (Repeat this activity a number of times so that the children can further clarify their thoughts.)
- Hold up the **apple shower gel**. "Who thinks this item is a 'yes'? Who thinks it is a 'no'? Who is 'undecided' about the item?" Tell the children that it is a "yes."
- Hold up the **pen**. "Who thinks this item is a 'yes'? Who thinks it is a 'no'? Who is 'undecided' about the item?" Tell the children that it is a "no."
- Hold up the **fresh flower**. "Who thinks this item is a 'yes'? Who thinks it is a 'no'? Who is 'undecided' about the item?" Tell children that it is a "yes."
- Finally, show the **nail clippers**. "Who thinks this item is a 'yes'? Who thinks it is a 'no'? Who is 'undecided' about the item?" Tell the children that it is a "no."
- Ask "Who thinks they have the idea?" Walk around the room one more time to determine children's thinking. When the majority or all of the children have the idea, tell them that on the count of three they are to say the idea together. "One, two, three..... Things that smell."

Debriefing:
- Ask if anyone came up with a different idea/concept that would also apply to the items in the activity. Discuss why the concept would or would not work.
- Discuss how everyone came to his or her conclusion about the concept. "What were some of the ideas you had while the items were shared? What items made you change your mind?"
- Take time to discuss the concept and clarify how certain items were indeed "yeses" or "nos."
- Ask children to share the feelings they experienced during the model.

SAMPLE ACTIVITY

Concept: Mammals
Definition: A mammal is any of a large class of warm-blooded, usually hairy, vertebrate whose offspring are fed with milk secreted by the mother.
Grade Level: 2nd-3rd
Source: Susie Coolidge, Anne Meyer, Jennifer Russ, and Katie Stratman (preservice teachers)

Words for Presentation (Symbolic level - words; Pictorial level - actual objects and/or pictures could be used with this activity):

Yeses (Examples)	Nos (Non-examples)
dog	coat
cat	daffodil
mouse	pumpkin
elephant	alligator
U.S. president	ladybug
whale	eagle
skunk	book
rabbit	tree
tiger	duck
you and me	frog

Order for Presentation: (If appropriate, depending on children's reaction to the activity, the presentation order may be adjusted.)

1. dog   2. cat   3. coat
4. daffodil   5. mouse   6. pumpkin
7. elephant   8. U.S. president   9. alligator
10. ladybug   11. eagle   12. whale
13. skunk   14. book   15. tree
16. rabbit   17. tiger   18. duck
19. you and me   20. frog

Presentation Format:
- Tell the children that you are thinking of an idea/concept that they will need to determine by carefully considering what attribute(s) all of the examples have in common that the non-examples do not have. Make two columns on the board, one for Examples and one for Non-examples.
- Remind the children that this is a silent model. They must respect each other's thinking space.
- Begin by writing the words **dog** and **cat** on the board under the word "Examples" and inform the children that these words fit your idea.
- Ask the children to begin a list (mentally or on paper) of any ideas they have about what both examples have in common.
- Ask the children by a show of hands, "How many of you have one idea? Two ideas? Three ideas? More than three ideas?"
- "Now, I will tell you two things that do not fit our idea/concept." Below the word "Non-example," write the words **coat** and **daffodil**.
- "Review your list of ideas. Check to see if everything still fits and cross off any ideas

that do not fit what the examples have in common that the non-examples do not have." Ask, "Did anyone cross off any ideas?"
- Inform children that you will now write a word on the board (not in either column) and you want them to think if the word is an example or a non-example of your idea. Write the word **mouse** on the board. By a show of hands ask, "How many think the word 'mouse' is an example? How many of you think it is a non-example? How many of you are undecided about where the word 'mouse' fits?" (Observe when the children raise their hands.)
- Erase the word 'mouse' and rewrite it under the Examples column and say, "Mouse is an example of my idea."
- Remind the children to carefully examine the Examples column and compare it to the Non-examples column. Ask, "How many of you have one idea? Two ideas? Three ideas? More than three ideas?"
- Write the word **pumpkin** on the board (not in either column). By a show of hands ask, "How many think the word 'pumpkin' is an example? How many of you think it is a non-example? How many are undecided about where the word 'pumpkin' fits?"
- Erase the word "pumpkin" and rewrite it under the Non-examples column. Say, "Pumpkin is a non-example. It does not fit my idea."
- Continue writing the words on the board, one at a time, asking the children whether they think each word is an example or a non-example of the idea, or if they are undecided about where the word fits. Then erase each word and rewrite it in the correct column, stating whether it is an example or a non-example of the idea.
- Periodically ask the children if they have one, two, three, or more thoughts about what the idea/concept is.
- Ask, "How many of you think that you have my idea?"
- Next, state, "I am going to walk around the room to see what you are thinking my idea/concept could be. Please hold up your hand if you want me to look at your idea. When I am next to you, either point to it on your paper or whisper it to me. I will not tell you whether or not you have my idea. I am checking to see what your thoughts are at this time." By checking the children's thoughts, the teacher can select words that might help those who are close to the idea narrow their thoughts. *Be flexible with the order of the words that you present.* You want your children to experience success.
- After checking for the children's ideas, ask for volunteers who think they know the idea to give examples. Write the volunteers' words on the board and ask, "How many think the word _____ is an example? How many think it is a non-example? How many of you are undecided about where the word _____ fits?" Erase the words and rewrite them in the correct columns.
- Ask again how many children think that they have the idea. Walk around the room to check on what they are thinking.
- When all of the children (or the majority) have the idea, ask children to say their idea together on the count of three. "One, two, three—Mammals!"

Debriefing:
- Ask the children if anyone has a different idea/concept that would also apply to the words in the activity. Discuss why the concept would or would not work. (Often, more than one idea will fit the examples and non-examples chosen.)
- Discuss how everyone came to his or her conclusion about the concept. "What were some of the ideas you had while the items were shared? What words made you change your mind?"
- Take time to discuss the concept and clarify why certain words were "yeses" or "nos."
- Ask children to share the feelings they experienced during the model.

Concept Attainment Variations

Variations of this concept attainment model can be developed and implemented on the early childhood level. These variations can be created for whole-class instruction, as learning centers for individual or group work, as interactive bulletin boards, and/or as activities to be sent home for parents to enjoy with their children. Adults and children can design their own mathematics and science concept attainment activities, thus increasing their enjoyment of such a conceptual learning experience. Parents can help their children create concept attainment activities to bring to school that will challenge their classmates.

In addition to early childhood mathematics and science concepts, social studies also includes many concepts that children can experience through engaging activities. Martorella (1998), in his book *Social Studies for Elementary Children*, recommends that educators strengthen conceptual development with elementary children by using various engaging formats, such as concept boxes, concept folders, and concept bulletin boards. Based on Martorella's suggestions, I recommend that the following concept attainment formats be created for early childhood learners.

Concept Mystery Boxes or Bags

Numerous options exist for using concept mystery boxes or bags with children. Children are naturally curious and enjoy acting like detectives by exploring clues and discovering information on their own; thus, they take ownership of their invented knowledge. Educators can use the following suggestions when designing mathematics and science conceptual activities with decorated mystery boxes or canvas bags. The activities can be used for whole-class instruction and/or in learning centers.

Variation One: Create a mystery box/bag by labeling every item in the container with either a "yes" or "no" tag. Children must separate all of the items, one by one, in the mystery box/bag and place them in a location labeled either "yes" or "no" ("thumbs up" or "thumbs down" also works). Children must determine what the "yeses" have in common that the "nos" do not have. Remind the children that they should be detectives and keep their idea(s) to themselves in order for everyone to have an opportunity to think on their own. When ready, each child can whisper his/her idea(s) to you or write his/her idea(s) on a small piece of paper that they slide into an Assessment box (a box with a slit in the lid). When all children have had an opportunity to share their idea(s), gather them together and share children's various responses. (This might be done within a few minutes or, if the mystery container is placed in a learning center, at the end of the day or week.) Children should have the opportunity to share why they chose their ideas. It is important to remember that more than one idea might appropriately apply to the content placed in the mystery container. Be open-minded and flexible with children's ideas. Remember to emphasize how exciting it is to learn about the different ideas that other people have. Each child's prior experiences with the items represented in the mystery container will vary, which could lead to formation of very different concepts/ideas.

Variation Two: Every item placed in the mystery container should be a "yes." It must fit the concept/idea. For example, if studying the color "blue," every item in the mystery box must have the color blue on it. Inform children that your mystery container has only "yes" items (examples) in it. They must be detectives and solve the secret mystery. Remind them to keep their ideas to themselves. This concept attainment activity can be used for whole-class instruction or placed in a learning center. "Take home" mystery containers can be created to involve parents/guardians in solving the mystery. In turn, with some parental assistance, children can return to school with their own mystery container so their classmates can act as detectives and solve the new mystery. Children can be assessed in a similar manner to that described in Variation One.

Variation Three: Create a mystery container with a mix of examples and non-examples of your concept. Tell the children your concept/idea (e.g., "Things on a Farm"), and have them separate the "yes" items from the "no" items. This concept attainment variation could be placed in a learning center. Children could work on the activity individually or in small groups. Paper, pencils, and colored markers should be provided in the learning center so that the children can draw a picture (or writers could print the words) for their final choices of examples and non-examples. The children could place their concept attainment drawings in a class file for completed work. After all the children have completed the concept attainment activity, have volunteers share their responses and explain the reasons for their choices. Children's projects could be posted on a bulletin board, or placed in individual portfolios.

Concept Folders/Large Business Envelopes

The following concept attainment variations can be placed in learning centers for the children to complete individually or in small groups. They also can be duplicated for children to complete as engaging concept attainment seatwork.

Variation One: Provide children with a large, yellow business envelope that contains a manila folder and a piece of poster board, or a laminated sheet of colored construction paper, with the words "Examples/Yeses" and "Non-examples/Nos" written at the top. Separate the heading words so that one half of the area is for examples and the other half is for non-examples. Also include laminated pictures and/or words of examples and non-examples of your concept. Inform the children of the concept in advance and have them place all of the laminated items under the appropriate headings. When completed, the children could ask an adult to review their work, or they could transfer their information to a piece of paper and place it in the completed class work file. After all the children have completed the concept attainment activity, gather as a class to review the children's responses and reasons for their choices.

Variation Two: Laminate pictures and/or words that represent examples/yeses and non-examples/nos of a chosen concept (e.g., objects that come in pairs) onto the inside of a manila folder. Inform the children in advance of the concept. With a dry erase marker or a grease pencil, have children circle the examples and place an "X" on the non-examples. When completed, children could ask an adult to review their work, or they could transfer their responses to an identical sheet of paper and place it in the completed class work file. After all the children have completed the concept attainment activity, review, as a class, their responses and reasons for their choices.

Note: With special thought, the same laminated folder could be used for more than one concept attainment activity. For example, after the children have completed the concept attainment activity on "objects that come in pairs," the teacher could inform the children that their next concept activity is to review the pictures/words again and this time circle the items that fit the concept "things that float." When completed, assess the children's work as previously described.

Variation Three: Similar to Variation One, provide children with a large, yellow business envelope that contains a manila folder and a piece of poster board, or a laminated sheet of colored construction paper, with the words "Examples/Yeses" and "Non-Examples/Nos" written at the top. Separate the heading words so that one half of the area is for examples and the other half is for non-examples. Also include laminated pictures and/or words that the children can use as examples and non-examples of a concept. *Do not* inform the children of a specific concept in advance. Instead, encourage them to carefully review all of the laminated items and separate the items into examples and non-examples of their chosen concept. Encourage the children to use all of the items. (Carefully choose items that lend themselves to a variety of concepts.) When completed, the children could ask an adult to review their work. The children also could transfer their final concept attainment activity to paper and place it in the completed class work file. After all the children have completed the concept attainment activity, meet as a class to review their various responses and reasons for their choices.

Note: This challenging variation also can be used with a mystery concept box or bag.

Concept Attainment Interactive Bulletin Boards

Variation One: At the top of a bulletin board, place a written or pictorial notice of the concept children will investigate, such as "things that fly." Make two columns on the bulletin board, one for Examples/Yeses and one for Non-examples/Nos. Attach a box or envelope at the bottom of the bulletin board, or place it on a nearby table. Fill it with laminated pictures and/or words that children will use on the bulletin board. Give the children a deadline for completing the activity (by the end of the day, week, etc.). One by one, the children choose an item from the container, without looking, then write their initials on the back of it with a dry erase marker or grease pencil and decide under which column, either "Examples/Yeses" or "Non-examples/Nos," it belongs. They then affix it to the appropriate section of the bulletin board. Be certain that the envelope or box contains enough items so that every child in the class will have an item to place on the bulletin board. When all items have been placed on the bulletin board, review, as a class, the children's choices and reasons for placing them where they did.

Suggestion: Because the children's responses are displayed publicly, allow them enough time to work on this project (at least a few days) and permit them to change their positioning. Those who place their items under the incorrect category usually rethink their decisions before the children are gathered as a class to discuss their choices. It is very important to allow children an opportunity to express why they chose to place their items where they did. Children who reconsider the placement of their items usually will share their thoughts on why they did so. Through this sharing, a variety of perspectives will be expressed, which extends thinking for both the children and their teacher concerning the concept investigated.

Variation Two: The teacher chooses a concept and arranges the appropriate pictures/words under the Examples and Non-Examples columns. The children are responsible for whispering their ideas to the teacher or sliding their responses into the assessment box. Inform the children in advance of the deadline for completing their responses. When all of the children have responded to the mystery concept, gather them as a class and allow them to share their various ideas and perspectives on the concept.

Variation Three (Challenging Extension to Variation Two): The teacher chooses a concept and arranges the pictures/words under the Examples and Non-Examples columns. Place a few items in an incorrect column, and challenge children to decide how all of the items would fit a mystery concept if they were rearranged. Inform them that a few items need to be moved to the other column, but *do not* inform them of the concept. Design the concept attainment activity so that more than one concept is possible, depending on how the children rearrange the columns. After deciding on a concept and correctly arranging the items, children should slide their responses into the assessment box. When everyone has had an opportunity to respond, gather the children and review the various concepts and arrangements.

Suggestion: While many children will be eager to tackle this discovery concept variation on their own, they also could work with a partner.

Sample concept attainment variations created by preservice and inservice teachers follow.

SAMPLE ACTIVITY
Mystery Box

Concept:	Things That Produce Sound
Concept:	Items that produce a noise when movement or air is applied.
Grade Level:	PK-2nd
Source:	Melanie Graham (preservice teacher)
Form of Presentation:	Objects will be hidden in a mystery box
List of items:	Bells, wooden rods, whistle, kazoo, red horn, wooden toy rattle, toy frogs (in a tin), orange hammer, noisemaker (Concrete level)

Description of Activity: (This activity could be used as an introductory lesson for a unit on sound)

- Begin the activity by explaining to children that the box contains items that fit your idea. Tell children that all of the items in the box are "yeses" and that they all have something in common (alike).
- Remind children that this is a silent activity, so that they have to keep their ideas to themselves.
- Reach into the box and pull out the **bells**. Tell the children that this item fits your idea.
- Tell the children to think of any ideas they have after seeing the bells.
- Reach in the box and pull out the **wooden rods**, and ask the children to think about what the bells and wooden rods have in common.
- Ask children if they have one idea, two ideas, or more than two ideas.
- Pull out the **whistle**, and ask the children to think how all three items are alike (the same).
- Pull the **kazoo** out of the box and remind children to think about what all of the items have in common. Hold up all of the items for children to see.
- Ask children if they have one idea, two ideas, or more than two ideas.
- Tell the children who think they know your idea to raise their hand and that you will walk around the room so they can whisper their idea to you. *Do not* tell children if their ideas are correct.
- Pull out the **red horn** from the box and remind the children to think about what all of the items have in common. Let the children come closer to look at the items, if they wish. Remind them not to put the items near their mouths.
- Pull the **wooden toy rattle** from the box and remind the children to think about what all of the items have in common. Tell them, "Think about how they all fit together."
- Ask the children to think of what the idea might be. Go around the room again and ask children to whisper in your ear what they think the idea is.
- Ask for a volunteer to share an item that they think would fit with the other items. Ask children if they think the item is a "yes" or a "no," or if they are "undecided" about the item. Ask for other volunteers to share items that they think would fit with the items presented, and then repeat the "yes," "no," and "undecided" questions.
- Pull out the **toy frogs (in a tin)**, then the **orange hammer**, and then the **noisemaker**. Ask children to think about how these items fit with the other items.
- Review all of the presented items. Be certain that everyone can see each item at all times.
- Ask children if anyone would like to get a closer look or "try out" any of the objects.

Remind the children that they can use only their hands (not their mouths). Also, remind children about appropriate behavior in advance.
- Ask children how many now think that they know your idea. Walk around the room and allow children to whisper in your ear what they think the idea is.
- When all or the majority of the children have the idea, have them say the idea together on the count of three.
- Ask children if anyone has another idea that would fit for all of the items.
- Allow children to share what their thoughts and feelings were as they progressed through the activity.
- Discuss the concept of items that produce sound, and demonstrate how each object does indeed make a sound. Children can demonstrate the hand-held items, while you demonstrate the items that produce sound when air is blown into them.

SAMPLE ACTIVITY
Mystery Basket

Concept:	Things That Tell Time
Definition:	Anything used to measure time in days, hours, minutes, seconds, or even a nonstandard measure of time
Grade Level:	K-3rd
Source:	Amy Dannis (inservice teacher)
Form of Presentation:	Objects will be hidden in a mystery basket
List of Items:	Clock, watch, pocket watch, timer, stopwatch, calendar, egg timer, planner, sundial, scoreboard, time capsule (Concrete level)

Description of Activity: (This is designed as an introduction to telling time. Show items one-by-one to the children using the following script)

- "I am thinking about a special idea. Your job is to figure it out, so put your thinking caps on. We have one important rule: You must keep your ideas to yourself. We want our friends to have a chance to figure out the idea on their own. At the end, I will let everyone share their ideas."
- "The exciting thing about this activity is that today I brought all 'yeses' in my magic basket. Every item fits with my idea."
- The teacher holds up the first item and reminds children that this is a "yes." Walk around and let everyone look at it.
- Continue this procedure with each item. After the second or third item is shared, ask, "Who has one idea? Two ideas? More than two ideas?"
- After sharing several items, allow children to whisper their ideas in your ear. Do not tell them if they have the correct idea. Use this information to choose the next items to share.
- Continue to share items and ask who has ideas. Share more familiar items so everyone will be able to figure out the idea.
- Allow several children to volunteer other items that they think will fit the idea.
- Walk around again and allow children to whisper their ideas to you.
- When everyone has the idea and all items have been shared, say, "On the count of three, use an inside voice to say what you think the idea is. One, two, three _____."
- Allow children to share other ideas that they had. Talk about how they figured the idea out. Was it one particular item that helped them figure out the mystery?
- Assessment will be based on the ideas children whisper and the discussion that follows children's choral response. The discussion will be helpful in assessing children's prior knowledge about telling time.

Extensions: The unit on time could continue after this introductory activity in the following ways:

- Have children bring something from home that can tell time
- Present a lesson about the sundial and how it works. Make a sundial.
- Make a class time capsule and vote on things to put in it.
- Time different activities using a stopwatch.

SAMPLE ACTIVITY
Mystery Box/Bag

Concept:	Things That Live Outside
Definition:	Plants and animals that normally live outside
Grade Level:	PK-3rd
Source:	Angela Tatum (preservice teacher)
Form of Presentation:	Mystery box/bag containing items (objects for the Concrete level and/or pictures for the Pictorial level) labeled "yes" or "no"

List of possible items:

Yeses	Nos
grass	heart-shaped container
leaves	rock
dandelions	candle
wild flowers	ball
twig	sunglasses
butterfly	book
worm	pencil
snake	paper
polar bear	jewelry
groundhog	shoe

Description of Activity: This activity can be placed in a learning center. Put all labeled items in a box or a bag and have children separate them into a "yes" group and a "no" group. Children should determine the concept based on what the yeses have in common that the nos do not have. Once the concept has been determined, children could whisper their idea to you or write their response on a piece of paper and place it in an assessment box. When all children have had an opportunity to visit the center, the teacher should gather the class together and discuss children's various ideas.

Extension: Children could collect additional "yes" items during outside recess time and then share why their item would fit the concept.

SAMPLE ACTIVITY
Poster/Envelope

Concept: Things Found in the Kitchen (Bathroom, Bedroom, etc.)
Definition: Items typically found in most kitchens
Grade Level: K-3rd
Source: Natalie K. Richards (preservice teacher)
Form of Presentation: Poster/Envelope and pictures/objects (Pictorial level)
List of Possible Items: Alarm clock, baking dishes, bed, canisters, casserole pan, chainsaw, chair, coffee pot, comfy chair, drill, electric shaver, food scale, garbage can, gas pump, hair dryer, hammer and wrench, hand mixer, iron, jigsaw, lamp, lawn mower, light bulb, mirror, mop, oven, paint roller, paper towels, plug, pots and pans, refrigerator, scale, shaving cream, shelf, shower head, silverware, table, television, toaster, toothpaste, toothbrush, utensils, vacuum cleaner, VCR, washing machine, water faucet, water hose, watering can

Description of activity: This poster/envelope activity is designed to be used with a unit on safety and placed in a learning center. Children will be given an opportunity to place various pictures/objects on either a "yes" poster or a "no" poster, according to the given concept. Each day, a different concept will be presented in connection with the unit on safety (e.g., things found in the kitchen, bathroom, bedroom, living room, etc.) In addition, appropriate safety procedures and cautions about the specific room will be discussed each day.

At this center, children will begin by looking at the concept for the day. Individually, or with a small group, children will decide on some objects that they think will fit the concept. Reaching into the envelope and pulling out a picture of a object, children are to decide whether they think that "yes," this object would be found in this room, or "no," this object would not usually be found in this room. Continuing with this procedure, children will review all of the pictures, placing them on one of the poster boards.

Note: Depending on the children's previous experiences or the various objects that they may have in the rooms of their house, their ideas may differ.

Assessment: Discuss with children their ideas while participating in the center activity. For example, ask, "Why do you think that this picture goes on this board?" Also, after everyone has had an opportunity to visit the center, the class could gather as a group and discuss their various thoughts concerning where they placed each item.

Extension: Children will draw pictures of other objects that they think will fit on the "yes" poster.

SAMPLE ACTIVITY
Manila Folder

Concept: Things Found on a Farm
Definition: Items typically found on most farms
Grade Level: K-2nd
Source: Tiffany Rohrer (preservice teacher)
Form of
Presentation: Manila folder and pictures/objects (Pictorial level)
List of Possible Items:

Yeses	Nos
chicken	policeman
cow	sailboat on water
hay	doctor
tractor	airplane
lamb	merry-go-round
farmer	giraffe
barn with a silo	train
horse	dinosaur
pigs	whale
barn with a farmer	beach

Description of Activity: Inside the top of a manila folder, write "Things Found on a Farm." Print the word "Yes" on one side of the fold and "No" on the other side. As an example for children to follow, tape or laminate one picture of a "yes" item and one picture of a "no" item under the correct headings.

This activity can be used in a learning center during a unit on farms/farm animals. To understand children's prior knowledge, it could be placed in a center at the start of the unit. In order for children to have many opportunities to experience the concept, it could be left in the center throughout the duration of the unit. To challenge children, different items could be placed in the folder during the week.

Assessment: On the outside of the folder, create a pocket with a picture of a farm on it. Inside the pocket, place a self-check sheet (laminated) that includes the correct placement of the items found inside the folder. Children should be informed that the self-check sheet is to be used only when they feel they have successfully completed the activity.

Extension: Have children draw a picture of a farm. Allow them to share their "yeses" with their classmates. Display their farm pictures for all to see and/or include them in children's portfolios.

Benefits of Concept Attainment for Children

Children must have an understanding of classification in order to participate in the attainment of concepts through discovery. Classification requires children to group objects/ideas together that have one or more common criteria. Children learn that objects can be grouped together using various common features/attributes. Early childhood mathematics and science contain many opportunities for children to group and sort objects according to categories. The activities for attaining concepts through classification found in this chapter provide both teachers and children with opportunities to become engaged in the inquiry process of learning, resulting in numerous cognitive and affective benefits.

Cognitive Benefits
- Develops observation skills, sensory skills, classification skills, critical thinking skills, decision-making and problem-solving skills, vocabulary and literacy skills
- Creates disequilibrium, which stimulates inductive reasoning and leads to a desire to make information fit (assimilation) or develop new schemes (accommodation)
- Reinforces previous knowledge by connecting new knowledge to prior knowledge
- Alters and improves concept-building strategies (Joyce et al., 2000)
- Provides opportunities for learning through presentation of contrary information, and helps to correct incorrect interpretations (misconceptions)
- Provides connections between subject areas
- Provides teachers with a conceptual alternative to induction, enabling them to control data sets and help children develop precise knowledge of concepts (Joyce et al., 2000)
- Allows teachers to determine whether ideas introduced earlier have been mastered (Joyce et al., 2000).

Affective Benefits
- Values individual experiences/perspectives and demonstrates an appreciation of diverse thinking
- Develops tolerance of ambiguity, yet an appreciation of logic (Joyce et al., 2000)
- Strengthens social skills within a group setting, as children learn patience and tolerance of self and others
- Motivates children to participate in an engaging, exciting, stimulating model that provides opportunities to solve mysteries/puzzles and reinforces one's natural desire to reach equilibrium.

Concepts for Attainment

The following list of various mathematics and science concepts can be developed into engaging concept attainment activities. With your imagination, the list can increase in size.

wetness	recyclable items
coldness	reusable items
dryness	things found in a desert
things that are round	things found in a rainforest
things that are square	things you can smell
things with points	springtime items
items with curves	winter items
items that multiply	things that have teeth

things to buy
things to sell
things that open and close w/Velcro
things that can be zipped
plants
animals
coastal states
states with mountains
states with deserts
things that run with power
science or math occupations
things made from metal
things made from wood
things made from plastic
things made from glass
things that float
things that sink
things found on a farm
things found in a city
items to tell time
graphs
things you can see
things you can hear
fractions or rational numbers
twoness, threeness, fourness, etc.
items that come in dozens
metric system
things in the kitchen, etc.
first-aid kit items
things that live outside
things with a shell
foods that must be kept cold
nocturnal animals
things that make noise
things with wings

things that we only have one of
things that come in pairs
items that reflect
items that absorb light
items that come from trees
items that vibrate
things that have feathers
things that have hair
parts of a rainbow
things that cause pollution
things that require air
items that have buoyancy
electricity
natural sky lights
items in an observatory
livebearers
things that give off heat
insects
amphibians
foods picked from trees
foods grown as roots
foods picked from vines
multiples of 2 (3, 4, 5, etc.)
parts of a whole
things with five
things with ten
vegetables
fruits
gardening items
things that feel soft
objects with numerals
bread/pasta/cereal food group
things with the color green, blue, etc.
symmetrical items
things with scales

Summary

We use concepts to organize our world. Jerome Bruner emphasizes the importance of learners attaining understanding of concepts through inductive reasoning. Learners of all ages should be challenged to successfully attain concepts through the use of carefully planned activities in which they utilize the process skills of observation and classification. Children need many opportunities to explore, observe, describe, select, estimate, and measure various items. Such experiences provide early childhood-age children with opportunities to classify, a fundamental skill for understanding whole number (Kellough et al., 1996).

Mathematics and science concepts can be acquired through the implementation of various concept attainment formats that are developmentally appropriate for children. Concept attainment variations can be presented for whole-class instruction, or placed in learning centers and/or on interactive bulletin boards. Children also can create their own concept attainment activities to challenge their classmates and family members. The list of mathematics and science concepts that can be experienced through such concept attainment models is vast. Although the concept attainment examples presented in this chapter were designed for learners of early childhood mathematics and science, the discovery and attainment of concepts following the described methods can be used effectively across all subject areas and at all grade levels.

References

Chaillé, C., & Britain, L. (1997). *The young child as a scientist: A constructivist approach to early childhood science education* (2nd ed.). New York: Longman.

Charlesworth, R., & Lind, K. K. (2003). *Math and science for children* (4th ed.). Albany, NY: Delmar.

Howe, A. C., & Jones, L. (1993). *Engaging children in science.* New York: Macmillan.

Joyce, B., Weil, M., & Calhoun, E. (2000). *Models of teaching.* Needham Heights, MA: Allyn and Bacon.

Kellough, R. D., Carin, A. A., Seefeldt, C., Barbour, N., & Souviney, R. J. (1996). *Integrating mathematics and science for kindergarten and primary children.* Englewood Cliffs, NJ: Prentice-Hall.

Martorella, P. H. (1998). *Social studies for elementary children: Developing young citizens* (2nd ed.). Upper Saddle River, NJ: Prentice-Hall.

Winnett, D. A., Rockwell, R. E., Sherwood, E. A., & Williams, R. A. (1996). *Discovery science: Explorations for the early years.* New York: Addison-Wesley.

Chapter Four
Concept Mapping for Early Childhood Learners

As children wonder about and experience the world around them, they interpret their environments and form ideas about the various phenomena they observe. Children apply the knowledge they accrued from experiences and have stored in long-term memory (schemes/schemata) to their new experiences (Kellough et al., 1996). While children are often correct in how they interpret phenomena, they also, of course, can be incorrect. As children mature, their interpretations of phenomena change; while young, children have many preconceptions and misconceptions about their world. Children often attempt to fit new experiences into those existing schemes, rather than extrapolate new or expanded knowledge from their newly acquired experiences.

An important task, then, for teachers is to design experiences that provide opportunities for children to correct their preconceptions and misconceptions. "The task of the teacher is to facilitate the learner's continuing accurate construction of old and new schemata" (Kellough et al., 1996, p. 13). Once a child believes a certain concept is correct, however, it can be challenging for educators to change the child's mind. Considering that children's beliefs influence their receptiveness to new information, educators are always eager to try instructional strategies that will help learners correct their misconceptions and accept accurate explanations of phenomena (Howe & Jones, 1993). Concept mapping can be an excellent tool for facilitating this assimilation and accommodation of knowledge, thereby helping children to change their misconceptions (Kellough et al., 1996).

Allowing children to work cooperatively while exploring their world is an excellent way to uncover and challenge misconceptions. As they work together, children discuss their experiences. Often, a child's peer will have a different perception of a particular experience. The children listen to each other, discuss their conceptions, and work toward a more accurate understanding of the phenomenon. Early childhood educators should design and facilitate experiences that help children acquire correct conceptions of their world. One method often used by science educators on the elementary and secondary level that facilitates accurate understandings of prior and new knowledge is called "concept mapping."

Joseph Novak, a science education researcher, was interested in strategies that help children learn how to learn (Howe & Jones, 1993; Novak & Gowen, 1984). He developed a concept mapping technique, basing it on David Ausubel's theory of meaningful learning (1963), whereby a learner links new concepts to previously experienced concepts. By clarifying connections between concepts, Novak's concept mapping technique has been successful in helping children change any misconceptions they may have about their world (Kellough et al., 1996).

A concept map is a graphic/visual representation of concepts that shows various relationships between concepts. A concept map serves as a tool for organizing knowledge, from the most general to the most specific. Using a hierarchical format, the more general concepts appear at the top of the map, while specific supporting concepts are at the bottom. For example, the general concept of a plant is related to the more specific concepts of leaves, stems, roots, flowers, air, water, sunshine, etc. (Daugs, 1993). When constructing a concept map, children write the word "plant" at the top of their maps and connect the other suggested concepts, which are written below. Children draw lines connecting the concept words and put in linking words (usually verbs) on the lines to show their perceived connections (see Figure 4). An arrow placed at the end of each connecting line signifies the flow of each idea, usually stated in short sentence form. In referring to the diagram, you will notice that the last word of one thought is the first word for the following connected thought. For example, "plants have leaves" would be followed by and connected to "leaves can be green."

Through the use of concept maps, children organize their thoughts in a graphic/visual format, while connecting concepts and linking prior knowledge to new knowledge. Concept maps provide children with opportunities to reflect on their own thinking as they consider what they know or what they need to have clarified. As children make these visual representations and share their knowledge, teachers can assess children's thinking and determine if their concept connections are accurate. When misconceptions are evident, those children with accurate conceptions often can clarify the misunderstandings within their learning groups. This process is supported by Vygotsky's theory of the zone of proximal development, whereby a more competent child or an adult facilitates accurate conceptual understandings. Another benefit of concept mapping is the development of language used to describe scientific and mathematical concepts (Kellough et al., 1996). As children gain understanding of concepts and their relationships, they increase their ability to learn additional concepts.

If a concept map demonstrates a misconception, educators can design hands-on, minds-on learning opportunities that allow children to experience for themselves accurate conceptions of phenomena that they previously misunderstood. Through the process of assimilation and accommodation, children can clarify their thinking and experience equilibration. Early childhood educators need to understand, however, that correcting children's misconceptions is not an easy task. It often takes time and patience. Educators must listen carefully to children and provide many opportunities for questions and discussion in connection with hands-on, minds-on learning.

Concept mapping provides children with opportunities to become actively involved in their learning. When constructing a concept map as a class, a "dynamic interplay between students and teacher and between students themselves" is created (Romance & Vitale, 1998, p. 76). When constructing concept maps in groups, multiple perspectives are considered and then shared as children express their individual versions about how specific concepts are related. Through concept mapping, children can solve problems; make decisions; apply knowledge; and reflect on, organize, and link to long-term memory what they have learned (Romance & Vitale, 1998). Related concepts become connected rather than fragmented.

Concept mapping, as currently represented in textbooks, is an appropriate technique for children who are readers and writers. At the early childhood level (ages 3-8), however, not all children are readers and/or writers. Those who are will be able to participate in creating graphic (symbolic) word representations for concept mapping. I propose that children who are not yet readers and/or writers also can participate in creating concept maps on their appropriate learning and developmental levels. This can be done through the use of objects (concrete level) and/or pictures (pictorial level), and by writing words on a board/chart or by manipulating word cards with Velcro or magnetic backing.

Charlesworth and Lind (2003) recommend that materials used for conceptual activities meet children's level of development. "For each concept included in the curriculum, materials should be sequenced from concrete to abstract and from three-dimensional (real objects), to two-dimensional (cutouts), to pictorial, to paper and pencil" (p. 35). Before introducing paper-and-pencil activities, teachers must provide children with opportunities to manipulate and move materials in order to understand the concepts. The following activities are examples of concept mapping techniques appropriate for children's various developmental levels, beginning with the concrete level (actions on objects), progressing to the pictorial/transitional level (pictures), and eventually concluding with the symbolic/abstract level (words).

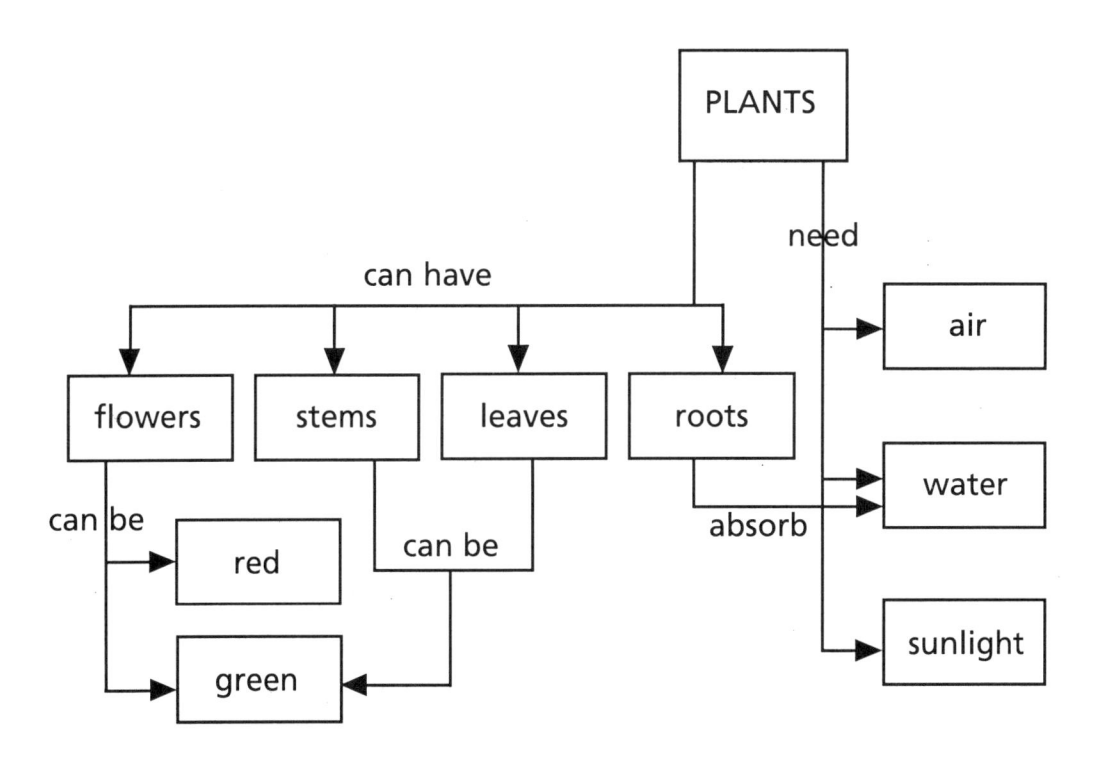

Figure 4

SAMPLE ACTIVITY
Concrete Concept Map

Topic/Theme: Animals
Grade Level: Preschool-2nd
Created by: Janell Bachelier (inservice teacher), Melanie Graham, and Natalie K. Richards (preservice teachers)
Materials: A large blanket for the floor, small/medium/large sentence strips for labels and arrows, large pictures of various living environments (e.g., water, land, sky), and a variety of beanbag animals.

Procedures: (Note: This project will be presented as a learning activity to follow a unit about the environment and animal habitats.)
- Begin by spreading out the blanket and asking the children to sit on the floor around the edges.
- Once the children are seated, scatter the various beanbag animals and the environment pictures on the blanket.
- Ask, "Looking at all of these things here, can you tell me what they all are?" (They're beanbag animals!) Ask, "What kind of beanbag animals are these? What do they represent?" (Different kinds of animals.)
- Display the sentence strip with the word "ANIMALS" (and a large picture with a variety of animals) on it at the very top of the blanket. Remind the children of your study unit: "Remember how we have been talking about different kinds of animals and where they might live? Well, today I'd like you to help me organize these animals and put them together (group them) according to where they might be found."
- Ask, "Who can tell me where one of these animals might live?" (A fish lives in water.) Place an arrow with its base at the word ANIMALS and point it toward the picture of "water." Write the word "lives in" on the arrow strip. Have a child who answered the question place the fish beanbag animal on the water picture.
- Continue with each of the other beanbag animals. Ask the children to share where each animal can be found (habitat). Children should place each beanbag animal in the appropriate habitat.
- Next, have children investigate the animals in each of the habitats and encourage them to create other groups by looking at the similar characteristics of the animals. Assist them with arrows and linking words.
- When all of the possible groups have been formed, ask for a volunteer to "tell the story" of the animals by following the arrows along the map. Model an example for the children first, so that they have an understanding of what you would like for them to share.

Extensions:
- Categorize the animal groups according to additional places where the animals might live, such as at the zoo, home, farm, etc.
- Focus on one habitat at a time, such as animals found on land, in water, or in the air.
- Children can string animals together and create an animal mobile that is also a concept map.
- Place animals and habitat words in a learning center for further investigation.

SAMPLE ACTIVITY
Concrete/Pictorial Concepts Map

Topic/Theme: Germs
Grade Level: Kindergarten-2nd
Created by: Aimee Jordan and Tiffany Rohrer (preservice teachers)
Materials: Towel, water bottle, tissues, cough drops, soap, a traced hand on a note card, Band-Aids, pictures of family, food, cut yarn strips, poster board, and note cards.

Procedures: (Note: This project will be used after a unit on germs has been presented.)
- Ask children to share what they have learned during the unit on germs.
- After the children are paired, distribute objects and their associated word cards (e.g., "water," "soap," "towel," "hands," "sneeze," "cough," "cut," "food," "Band-Aid," etc.) to each pair of children.
- Give the children time to observe their object with a partner and talk about how the object relates to germs.
- The children then share with the class what their object is and how it relates to germs.
- Explain to the children that they will make a word map showing how the objects and the words on their cards relate to germs.
- Guide the children by listening to their suggestions and giving clues when needed (e.g., giving a linking word such as "spread by" and letting children figure out who has objects that spread germs).
- The children can make a physical concept map by lining up in groups, and then using yarn and linking words to connect their objects/words to the word "germ."
- When completed, children place their objects/words and yarn on the floor and observe their creations.
- Children then explain their concept map on germs and tell how and why the yarn connects the objects.
- Encourage children to think of other ways that the objects/words might be connected.
- Listen to the children's explanations in order to evaluate if they connected the objects/words correctly. Record their accomplishments on a checklist.

Extension: Place concept mapping materials in a learning center where the children can explore further connections between the objects/words.

SAMPLE ACTIVITY
Concrete/Pictorial Concepts Map

Topic/Theme: Food Groups/Balanced Meals
Grade Level(s): Preschool-1st
Created by: Patty Stone (inservice teacher)
Materials: Play food representing the four food groups (dairy, meat, fruits/vegetables, grains) as well as sweets/junk food; picture cards representing the four food groups and sweets/junk food; paper plates; yarn, string, or crepe paper for concept links; linking cards with words "can have"/"can't have," "will have"/"will not have," "can include," "such as," etc.; stuffed Cookie Monster; lunch box (empty).

Procedures: (Note: This lesson will be used as one way of evaluating what the children have learned following lessons about the food groups and balanced meals.)
Teacher's role: Engage the children by holding up the Cookie Monster and asking them what it is. Ask them to remember what the Cookie Monster likes to eat. Initiate a discussion on the Cookie Monster's eating habits and ask the children what foods would be more appropriate for him to eat. Then hold up an empty lunch box and ask the children as a class to sort through the bucket of play food items to help Cookie Monster pack a healthful lunch. The picture food cards, yarn/crepe paper, and paper plates will be available as a way of sorting the play food items into categories and making accurate connections for a concept map.
Children's role: With the teacher's guidance, the children sort through the picture food cards and discuss which food groups are healthful or not healthful. Food cards should be placed in a horizontal row on the floor, or on a large table, below the Cookie Monster and his empty lunch box. (Note: Place items where *all* the children can be participants in the concept mapping activity.) The children then discuss whether each play food item is healthful or not. Paper plates should be placed below the food cards (one plate per food card) so that the children can place the play food on the appropriate food card category (e.g., all fruits/vegetables will be placed on the plate below the fruits/vegetables food card; all sweets/junk food will be placed under the sweets/junk food card, etc.).
Extensions: Children then pack different balanced and healthful meals for the Cookie Monster by placing an assortment of play food items in the empty lunch box. Separate meals can be packed for breakfast, lunch, and dinner. Children can be asked what food items not present could be added. For reinforcement, this activity can be placed in an interest or learning center after being introduced during whole-class instruction.

Note: An elaborate concept map can be used to direct learning for the school year or for higher grade levels. Children can draw upon previously learned concepts while making connections.

SAMPLE ACTIVITY
Pictorial Concept Map

Topic/Theme: Five Senses
Grade Level: Preschool-1st
Created by: Pam Jordan (inservice teacher)
Materials: Large picture of a young child with lines drawn to an eye, an ear, the mouth, the nose, and a hand; five senses cards with pictures and words for each sense; pictures of items (or actual objects) that children like to taste, touch, smell, see, and hear; cardboard arrows; foam mapping board with Velcro (can also use a flannel board or a magnetic board).

Procedures:
- While holding up the picture of a child, and pointing to the lines drawn to the specific body parts, ask children to think about what they recently discussed in science about their bodies. The children will volunteer, "The five senses."
- Place the picture of the child at the top of the concept board.
- Inform children that you are going to place, below the picture, a strip of paper that reads "Body parts are."
- Ask children to think about the body parts that we use for our five senses. Inform them that an arrow (hold the arrow up for all to see) will be used to point from the picture of the child to each of the body parts that they mention.
- Ask for volunteers to state the name of one body part that we use for our five senses (eyes, ears, nose, mouth, or hands). One by one, have each child find the picture card with the word that matches the body part that he or she shared. Have each child place their picture card horizontally below the picture of the child, using arrows (vertically) to "map the path" from the picture of the child to each body part.
- One by one, ask the children to think about what we do with our eyes, mouth, ears, etc.
- Have volunteers share that we see with our eyes, we hear with our ears, etc. Each volunteer should find the picture/word card that states the actual sense and place it below the body part, using an arrow to point from the one to the other.
- Ask the children to think of things that they like to hear, and have them share their thoughts. Have the children find a picture word card that would fit the sense of hearing and place it below the picture word card for "hear." Have each child connect his or her sense to the picture that fits what he/she likes to experience with the specific sense. For example, place an arrow on the board pointing from the picture of the ear (hear) to the picture of a child listening to music. Additional things to hear can be placed under the picture word card for the sense of hearing.
- Continue the above process for all of the senses by asking children to connect what they like to do with each sense.

Extensions:
- Place the concept board in an interest or learning center where children can practice connecting related concepts about the five senses.
- Children can draw pictures of what they like to do with their five senses and add them to the concept map.
- Children can bring objects or pictures from home to add to the concept map.

SAMPLE ACTIVITY
Pictorial Concept Map

Topic/Theme: Animal Classification
Grade Level: 2nd-3rd
Created by: Katie Fitzpatrick and Brandi Lamb (preservice teachers)
Materials: Pictures of animals; pictures of walking, flying, land, water, air; arrows; scissors; glue; construction paper; magnets.

Procedures:
- Fill the room with animals (stuffed, live, and photographed) from three classifications, which will be discussed and used as children review animal classifications.
- Encourage the children to recall the scientific discoveries they have made.
- Say, "I am now going to display several pictures and words that we have not classified. Look at these words/pictures and raise your hand when you can tell me a general category that all of these words/pictures might be placed under." (animals)
- Once a general category has been named, explain that the children should work toward more specificity as each new word/picture is placed beneath the other, in order to show the connections/relationship among them. Also explain how they will need to link each word/picture to the other words/pictures by selecting a linking word/phrase.
- Give examples of linking words that should be included, such as "can be" or "move by," etc. At this time, ask the children if they have any suggestions for linking words.
- Seek a volunteer to suggest what categories can be directly linked beneath "Animals," and how they would link those pictures/words to "Animals." Children may suggest that "mammals, amphibians, and birds" fit, because they are all types of animals.
- Encourage other volunteers to branch off from each previous idea, using only the words/pictures available to them, until the entire map is completed and all of the words/pictures provided have been used. (This can be done as a large group if all class members can see the words/pictures well; otherwise, it should be done in small groups.)
- With the children's assistance, read the final map out loud in sentence format, beginning at the top. Ask who agrees that the map makes sense the way it was read, who disagrees, and who is unsure.
- Direct the children to think of other ways the map could be constructed. Inform them that each child will create his or her own concept map at their seats. (Children can choose to work with partners or by themselves.)
- Ask for volunteers to share their concept maps by manipulating the magnetic words/pictures at the board, and then reading their maps and explaining their thoughts.
- After each map is presented, ask the other children to either accept or reject the new map. For a map that is determined to be unacceptable, children will problem solve and discuss why they rejected it. Be sure to emphasize that as long as the map presents clear, logical, accurate conceptions, it is acceptable. Explain that thought processes can differ, and that all viewpoints should be respected.
- Display the children's completed concept maps on a large picture of a brain to demonstrate how our brains work differently. The children should have an opportunity to read their classmates' maps and learn how others organized the information.

Extension: Place the concept mapping materials in a learning center where the children can expand on their experiences mapping the topic of animal classifications.

SAMPLE ACTIVITY
Symbolic Concept Map

Topic/Theme: Shelter/House
Grade Level: 2nd-3rd
Created by: Sara Lipner and Megan Weil (preservice teachers)
Materials: Ten laminated words with magnetic strips attached to the back (house, igloo, teepee, log cabin, Eskimos, pioneers, Native Americans, ice, canvas, wood); magnetic board, chalk or dry erase markers; paper and pencils.

Procedures: (Note: This concept map can be presented to the children before or after a unit on shelters (houses). Both methods will test children's prior or gained knowledge.)
- Begin by randomly placing the 10 words on the magnetic chalkboard.
- Ask the children what connections they can see among these words. The children could determine that they are types of shelter, and the materials for making the shelters, for the different types of people.
- Once the children have decided that all 10 words can be connected, they will be asked to create a concept map using these words. Ask the children which word they think should be placed at the top of the map. (House)
- The children then create their concept maps individually, with a partner, or in a small group. Remind them to include connecting lines, linking words, and arrows. The children should create as many concept maps as they can, using each of the 10 words only once in each map.
- After the children have created their concept maps at their seats, ask them to re-create them on the chalkboard. They will use the magnetic words and chalk for writing linking words on the connecting lines with arrows. The children who volunteer to re-create their maps on the board will read and explain them using complete sentences. The class then will discuss why they think each map is accurate, or if there is something they do not understand or agree with. Emphasize the importance of valuing diverse thinking as the children present their concept maps.

Extension: Set a learning center up at one end of the magnetic board for further experiments with the concept map.

SAMPLE ACTIVITY
Symbolic Concept Map

Topic/Theme: The Human Body
Grade Level: 2nd-3rd
Created by: Anna Haas, Amber Lakes, and Jerri Schriefer (preservice teachers)
Materials: Magnetic index cards with the following words printed on them (body, organs, hands, fingers, toes, heart, lungs, feet, muscles, bones, skull, arms, legs, brain, skin, biceps); magnetic blackboard or another large magnetic surface; chalk to draw arrows and linking words between the various cards.

Procedures: (Note: Ideally, this activity would be incorporated into a unit on the human body. It could be used as a precursor or introduction to a unit on the human body, as a concluding assessment measure, or somewhere in-between. It is flexible in nature.)
- Divide the children into pairs for cooperative group work. Give each group a list of the concept words and review the words as a class to ensure that all children know what each word is and means.
- Review with the children how to create concept maps. A concept map should be hierarchical, with the most general concept placed at the top and specific concepts placed at the bottom.
- Allow children adequate time to create their concept maps with paper and pencil. Concept maps should incorporate all of the given concept words. (Option: Provide children with sets of concept cards to manipulate prior to drawing their maps. Non English-speakers should be provided with picture/word cards.)
- After children have finished creating their concept maps, ask for volunteer groups to share their maps. Volunteer groups should go to the magnetic board and manipulate the concept word index cards to their liking. Chalk or dry erase markers will be provided for children to draw the corresponding lines, arrows, and linking words.
- As a class, discuss and compare children's concept maps. Are the concept maps the same or different? Is there consensus among the groups? Why or why not?

Extension: Place the concept mapping materials about the human body in a learning center with a magnetic board so that the children can continue to experiment with various connections among the concepts.

Benefits of Concept Mapping Activities

Both children and their teachers benefit when concept mapping is incorporated into the early childhood curriculum. Some of these benefits follow:

Benefits for Children
- Provides opportunities to see logical connections between new material and previously acquired knowledge, while organizing thoughts from general to specific, resulting in meaningful learning
- Improves children's concept constructions, while addressing preconceptions and correcting misconceptions
- Addresses all learning modalities: visual, auditory, tactile/kinesthetic
- Emphasizes diverse perspectives through the acceptance of various formats for use with the same topic, showing children that there are many correct ways to process similar information
- Provides active learning opportunities for children to organize their thoughts in a concrete/visual format through grouping and sorting, while manipulating concepts mentally and physically
- Promotes critical thinking skills—observation, comparison, classification—as well as problem-solving and decision-making skills
- Reinforces and strengthens communication, language, and literacy skills
- Promotes and values both cooperative learning and independent work
- Motivates learning while entertaining.

Benefits for Teachers
- Can be used to introduce or conclude a topic, as an activity within a lesson/unit, or as an organizer for units/lessons/activities
- Can be used as a pre- and/or post-assessment instrument for analyzing children's thinking—both formally and informally
- Allows teachers to observe gaps in children's knowledge in order to facilitate correct conceptions (connections)
- Encourages educators to become more open-minded and flexible about children's various interpretations and perspectives
- Provides for various learning formats: whole-class instruction, learning center, or interactive bulletin board
- Provides opportunities for subject integration.

Summary

Concept mapping offers opportunities for children to become actively involved in their learning while linking knowledge to long-term memory. When using concept maps, children organize their thoughts in a concrete and/or graphic/visual format, while connecting concepts and linking prior knowledge to new knowledge. Related concepts become connected rather than fragmented. Concept maps also provide children with opportunities to think about their own thinking as they reflect on their conceptual understandings. In addition, teachers can use concept mapping as an effective learning tool for assessing early childhood learners' understandings through their creation of concrete and/or graphic/visual representations. Through concept mapping, teachers can assess their students' concept connections and determine the steps necessary for further learning.

References

Ausubel, D. P. (1963). *The psychology of meaningful verbal learning*. New York: Grune & Stratton.

Charlesworth, R., & Lind, K. K. (2003). *Math and science for children* (4th ed.). Albany, NY: Delmar.

Daugs, D. R. (1993). *Improving science teaching*. Logan, UT: The Home Teacher.

Howe, A. C., & Jones, L. (1993). *Engaging children in science*. New York: Macmillan Publishing.

Kellough, R. D., Carin, A. A., Seefeldt, C., Barbour, N., & Souviney, R. J. (1996). *Integrating mathematics and science for kindergarten and primary children*. Englewood Cliffs, NJ: Prentice-Hall.

Novak, J., & Gowen, D. B. (1984). *Learning how to learn*. Boston, MA: Cambridge University Press.

Romance, N. R., & Vitale, M. R. (1998). Concept mapping as a tool for learning: Broadening the framework for child-centered instruction. *College Teaching, 47*(2), 74-79.

Chapter Five
Tying It Together

Concepts are acquired when they are understood. When children are actively involved in learning, they are more likely to understand concepts than when presented with isolated facts. Early childhood-age children must be active participants in learning process skills and fundamental math and science concepts. The concepts acquired at this level will be applied later when exploring and discovering more abstract concepts in mathematics and science. For children to reach their full potential, educators must design effective learning experiences that emphasize critical thinking, problem solving, and decision-making through active engagement. Children at all levels should have opportunities to develop and expand their critical thinking skills while creatively constructing mathematics and science concepts.

Many early childhood theorists promote hands-on, minds-on discovery learning. Various critical thinking teaching/learning models are effective for a wide range of ages, and can be adapted to suit early childhood-age children. The models presented in this book that promote critical thinking and discovery learning are the 5E Learning Cycle, Concept Attainment formats, and Concept Mapping.

Review of Presented Models

The first model presented was the 5E Learning Cycle. The Learning Cycle model was fueled by Piaget's constructivist theory and credited to Robert Karplus and his team of educators in the 1960s. While this model initially was introduced in the Science Curriculum Improvement Study (SCIS) program, many educators now incorporate it across the curriculum. In terms of children's retention of concepts, it is a highly effective learning model that allows children to take ownership of newly constructed concepts. In the 5E Learning Cycle variation, children have opportunities over five stages to build on prior knowledge; discover and reinvent concepts on their own; and tie concepts to relevant knowledge, career awareness, and Science, Technology, & Society.

The second set of presented models are inspired by Jerome Bruner's focus on attaining concepts through critical thinking and discovery. Through these concept attainment activities, children have opportunities to draw on prior knowledge as they compare and contrast information provided by the teacher. Children act as detectives as they mentally and physically categorize information inductively in order to reach specific concepts. Teachers can encourage and strengthen a respect for diverse thinking as children share their thinking experiences at the conclusion of each model.

The third presented model is based on Joseph Novak's Concept Mapping technique (1993), which focuses on identifying connections between and among concepts. Concept mapping is supported by Ausubel's (1986) theory that stresses the importance of tying prior knowledge to new concepts. An emphasis also is placed on the value of organizing knowledge in a graphic/visual format. Children's misconceptions about mathematics and science concepts can be addressed through the use of this model.

Benefits of Presented Models

These learner-centered models address all learning modalities—visual, auditory, and tactile/kinesthetic. While participating in these models, children develop observation skills, sensory skills, classification skills, critical thinking skills, literacy skills, and decision-making and problem-solving skills. They are challenged to build on prior knowledge and extend their thinking as they actively explore and reinvent concepts, while taking ownership of newly acquired knowledge. Learning occurs when children experience disequilibrium and respond by striving to reach equilibration through the assimilation and accommodation of knowledge. In doing so, opportunities exist for students to correct preconceptions and misconceptions through the use of process and critical thinking skills. Children also have opportunities to reflect on their own thinking processes as they consider their conceptual understandings. Furthermore, the models allow children to perceive concept relevancy, thus making learning more meaningful. By identifying connections between and among subject areas, children gain deeper understandings of content as concepts become connected, rather than segmented. Affectively, the presented models capitalize on children's natural sense of curiosity. As a result, learning is solidified and extended, which strengthens children's sense of efficacy with both content knowledge and process skills. Children also benefit socially as they learn to work cooperatively with others and value their peers' various perspectives. Furthermore, children are evaluated more equitably, through alternative means rather than only pencil-and-paper assessment tools.

Summary and Conclusion

These effective teaching models emphasize the importance of critical thinking for early childhood learners in mathematics and science. The various engaging teaching/learning models outlined in this book promote problem-solving, decision-making, and cooperative learning skills through the creative construction of mathematics and science knowledge, and they are linked to developmentally appropriate practice.

To reiterate, early childhood educators must value their role in designing effective learning environments that foster critical thinking. Teachers play an important role in creating well-informed contributing citizens who can think creatively and critically, and who can solve problems and make effective decisions.

The models presented in this book are meant to address both the NCTM and national science curriculum standards for teaching mathematics and science at the early childhood level. The checklist in Figure 5 offers strategies for effectively teaching early childhood mathematics and science.

In addition to creating stimulating learning environments, the 5E Learning Cycle, Concept Attainment models, and Concept Mapping offer numerous opportunities to conduct action research with children. Witness your students' enthusiasm as they share their discovered mathematics and science knowledge with you and their classmates. I wish for you and your children many enjoyable teaching and learning experiences in reinventing mathematics and science knowledge!

Recommendations for Creating Effective Mathematics and Science Learning Environments

- Allot time for children to experience free exploration and play with math and science learning resources.
- Design activities and lessons that support children's natural way of learning through the active construction of knowledge.
- Allow children to assist in the design of integrated thematic units based on their interests.
- Build upon children's knowledge by connecting their informal knowledge/ strategies to formal knowledge/strategies.
- Respect children's levels of learning by providing concrete experiences before introducing symbols.
- Provide a variety of activities that reach children's various learning modalities and intelligences.
- Use thought-provoking questions or problems to motivate interest in a particular topic or to introduce new concepts.
- Display enthusiasm and interest when teaching math and science in order to prevent math and science anxiety; instead, focus on building children's levels of efficacy.
- Encourage interaction through cooperative groups, as well as through peer and cross-age tutors.
- Incorporate challenging activities/games/puzzles into the curriculum that are thought-provoking, exciting, and include an element of chance.
- Design activities that are at children's developmental levels (concrete, pictorial, or symbolic) and that will allow them to experience success.
- Connect math and science content to real-life situations, STS (Science, Technology, & Society), and career opportunities to demonstrate relevancy.
- Encourage connections among the various content areas through subject integration.
- Focus on the language and communication of ideas in early childhood by linking math and science to literacy activities and children's trade books.
- Promote intellectual autonomy by providing children with opportunities to become more self-directed.
- Incorporate alternative assessment approaches into the curriculum for a truer picture of children's actual mathematical and science abilities.

Figure 5